RADICAL WINGS
& WIND TUNNELS
Advanced Concepts Tested at NASA Langley

Joseph R. Chambers & Mark A. Chambers

specialtypress
PUBLISHERS AND WHOLESALERS

Edited by Nicholas A. Veronico
Designed by Monica Bahr

ISBN-13   978-1-58007-116-1

Item #SP116

39966 Grand Avenue
North Branch, MN 55056 USA
(651) 277-1400 or (800) 895-4585
www.specialtypress.com

Printed in China

Chambers, Joseph R.
   Radical wings & wind tunnels : advanced concepts tested at the NASA Langley / Joseph R. Chambers & Mark A. Chambers.
      p. cm.
   Includes index.
   ISBN 978-1-58007-116-1
   1. Langley Research Center--History. 2. Aeronautics--Research--United States--History. 3. Research aircraft--United States.   I. Chambers, Mark A. II. Title.
   TL521.312.C442 2008
   629.1072'47555412--dc22
                   2008025090

Cover:
The Boeing Phantom Works X-48B Blended Wing Body research aircraft is mounted for wind-tunnel tests in the Langley Full Scale Wind Tunnel in 2006. The 21-foot wingspan vehicle subsequently served as a backup to a second X-48B during flight tests at the NASA Dryden Flight Research Center. The X-48B is a cooperative venture between The Boeing Company, NASA, and the U.S. Air Force Research Laboratory. (Photo courtesy of The Boeing Company)

Title Page:
The Lippisch DM-1 glider mounted in the full-scale tunnel for aerodynamic testing. Test program focused on the high-lift, low-speed aerodynamic characteristics of the delta-wing configuration. The maximum lift of the DM-1 was found to be much lower than values obtained for generic sub-scale models in the United States, leading to an early study of the control of vortex flows. (Photo courtesy of NASA)

Back Cover, Top:
Langley technician Joe Denn prepares a large powered free-flight model of the Hiller X-18 tilt-wing experimental aircraft for tests to measure its aerodynamic characteristics in the full-scale tunnel in 1957. The model used dual counter-rotating propellers to simulate the full-scale aircraft propulsion system, and was powered for these tests by electric motors with power routed up the model test apparatus. So-called "force tests" using internally mounted electric strain-gauge balances are routinely conducted to document and analyze aerodynamics before free-flight tests are started. (Photo courtesy of NASA)

Back Cover, left center:
Langley aerodynamicists joined a team from General Dynamics to design an efficient supersonic wing for the F-16. Using design methods developed and validated by Langley in its supersonic transport studies, the participants defined several candidate wing planforms that would meet supersonic cruise requirements. One of the study requirements was to maintain a transonic maneuver capability equal to or better than the baseline F-16. Shown here in the Langley 7- by 10-Foot Tunnel in 1978 is a model used to provide validation data for a computational study of the impact of camber and twist on transonic performance. The study's results proved that a supersonic-wing design could have outstanding transonic maneuverability and influenced the final shaping of the wing of the F-16XL prototypes. (Photo courtesy of NASA)

Back Cover, right center:
One of Langley's areas of expertise is aeroelasticity, including the prediction and control of airframe flutter. As the F-16XL program progressed, the configuration became a candidate for an advanced strike fighter for the Air Force and was redesignated the F-16E. This photograph shows a large model of the F-16E mounted in the Langley 16-Foot Transonic Dynamics Tunnel in late 1981 during flutter-clearance tests. Note the instrumentation on the vertical tail used to measure dynamic tail loads. (Photo courtesy of NASA)

Back Cover, bottom:
Langley's F-106B vortex-flap research aircraft in flight in 1987. Note the flow-visualization equipment housing on the turtleback of the aircraft. Flow over the left wing panel was illuminated during flights at night. The right wing panel carried pressure instrumentation and other devices to measure flow properties. The modified F-106B's subsonic and transonic performance was dramatically better than the baseline airplane. (Photo courtesy of NASA)

TABLE OF CONTENTS

ACKNOWLEDGMENTS

The authors would like to thank many individuals and our long-time associates for their outstanding assistance in the production of this book. Special thanks go to Patricia A. West and Sue B. Grafton of the NASA Langley Research Center for their personal efforts to preserve photographs and technical reports of the aeronautical history of Langley, and for providing their interest and encouragement in this project. A.G. "Gary" Price, retired head of the NASA Langley Research Center Office of External Affairs, provided access to photo histories prepared by the Langley Office of Public Affairs. Garland Gouger and the staff of the Langley Technical Library assisted in locating many rare technical reports and documents. Alicia Bagby and her associates at the Langley Photographic Archives Office also deserve special accolades for their outstanding responses to requests and historical information. Graphics support was kindly provided by Lee Pollard. Historical information for technical projects was graciously contributed by active and retired Langley researchers Joseph L. Johnson, Jr., Del Croom, Dan Vicroy, Mike Fremaux, and Mark Croom.

In addition to these wonderful Langley peers, we express our deep gratitude to Jennifer MacLean and the staff of the Air Zoo, Kalamazoo, Michigan, for permission to use photographs of the restored XP-55 aircraft and to Charles L. Day for his valuable insight and photographs of the XFG-1 fuel glider project. We also thank David Everett for permission to use his photographs of his father's flight-test activities in the Flex-Wing, Fleep, and XV-5A projects. Eric Hehs of Lockheed-Martin was especially responsive to requests for photographs of Convair and Lockheed projects. Sincere thanks go to Mary E. Kane, Thomas J. Koehler, Henry T. Brownlee, Jr., and Thomas J. Lubbesmeyer of The Boeing Company for their assistance in obtaining archival photographs. Special recognition also goes to Nicholas Veronico, Molly Koecher, and the staff of Specialty Press for their support of this endeavor.

This book is dedicated to the thousands of current and retired employees of the NASA Langley Research Center. Their efforts began with research on early biplanes that literally staggered through the air, and led the way through the evolution of powered flight to the mind-numbing achievements of today's advanced aerospace vehicles. The legacy of Langley's contributions to most of the critical aeronautical achievements of this nation is a direct result of their innovation, dedication, and expertise. The mission of the authors is to reveal a sample of those unknown efforts to the world.

Joseph R. Chambers
Mark A. Chambers
Yorktown, Virginia

THE ROLE OF NASA IN ADVANCED AERONAUTICAL RESEARCH

The reputation of the National Aeronautics and Space Administration (NASA) as the Nation's "Space Agency" has dominated the public's perspective of its contributions and scientific value in recent years. Almost constant publicity and media coverage of NASA's efforts in manned space exploration have naturally resulted in a widespread lack of awareness of NASA activities and considerable contributions in other critical fields, such as atmospheric sciences and aeronautics. More importantly, with every passing day much of the corporate knowledge and documentation of these less-visible efforts is rapidly disappearing as individuals pass away and the agency's focus moves away from aeronautics.

In the science of aeronautics, NASA and its predecessor—the National Advisory Committee for Aeronautics (NACA)—have played key roles in the evolution of manned flight and its benefits for the civil and military communities. Many legendary research and development examples have been documented to testify to the value of these efforts in aeronautics, which have resulted in significant advances in aircraft performance,

safety, and operational capability. The research staffs and facilities of the NASA Langley Research Center at Hampton, Virginia; the NASA Ames Research Center at Mountain View, California; the NASA Dryden Flight Research Center at Edwards Air Force Base, California; and the NASA Glenn Research Center at Cleveland, Ohio, have all contributed to this rich heritage of aeronautical experience. Well-known examples of contributions at Langley include the development of the NACA cowling for radial engines, laminar flow airfoils used by the P-51 Mustang in World War II, the X-1 research aircraft, the X-15 hypersonic research vehicle, variable-sweep wings, composite materials now used extensively in civil and military aircraft, the area rule used by many high-performance military aircraft, and the winglets and supercritical airfoils used by commercial transports and business jets. Most of these accomplishments have been publicized and documented extensively for the aviation enthusiast and historian.

However, for every successful advance in aeronautical technology, literally thousands of revolutionary aircraft concepts and ideas were conceived, assessed, and subsequently discarded for a multitude of reasons, including unsolvable technical barriers, lack of practicality, unacceptable impacts on the environment, excessive development or operational costs, or lack of a mission or market. In their role of providing the leading-edge aeronautical research building blocks used by industry to design and develop radical new aircraft, NACA and NASA have rich histories of involvement in revolutionary aircraft concepts. Many of these concepts were abandoned or stymied by the lack of technology and are among the lesser-known aeronautical efforts in aviation history. In most cases, the public is unaware that these government agencies were even involved in the projects.

The purpose of this book is to discuss some of the more interesting radical aircraft concepts that have been studied at Langley and to provide documentation of the research efforts and results that emerged from the studies. All of the aircraft concepts discussed were prototypes or concepts for experimental aircraft that did not, or have not yet, proceeded into production. An in-depth discussion of the thousands of relevant NASA studies of this nature at all four aeronautical research centers would require several volumes and be far beyond the intended scope of this book. Instead, this material focuses on interesting research activities conducted in a few select facilities at Langley and is limited to studies of fixed-wing aircraft. The aircraft projects were chosen based on either the unconventional aircraft configurations involved or the projected advantages of their revolutionary capabilities. Many of the examples presented herein involve famous industry organizations and historical figures in this nation's aeronautical history.

The existing technical documentation of activities related to many of the topics discussed is very limited. Many of the projects produced highly classified reports at the time of the studies and were immediately transmitted to a few recipients within the military and industry. In addition to reviewing the original documents, the authors obtained photographs from the Langley archives, personal collections, and other individuals that illustrate activities for the projects. Many of the photographs have never been published before.

In order to provide the reader with a helpful background and understanding of the interactions between NACA, NASA, and their aeronautical customers, a brief overview follows describing the charter, research capabilities, and contributions of the NASA Langley Research Center.

NASA Langley Research Center

Located in Hampton, Virginia, near the mouth of the Chesapeake Bay, the NASA Langley Research Center is the oldest NASA center, and in recognition of that fact is frequently referred to as the "Mother Center." Founded as

View of the West Area of the NASA Langley Research Center in Hampton, Virginia. Home to about 3,300 civil-service and contractor employees, the center conducts aeronautical and space-related research. The West Area was initially constructed in the early 1940s and now includes a myriad of facilities, including the Langley flight hangar in the foreground, numerous wind tunnels, and research laboratories. The Lunar Landing Research Facility gantry seen in the distance was put into operation in the 1960s to train astronauts for the Apollo moon landings. After Apollo, it was adapted for crashworthiness testing of full-scale aircraft and rotorcraft. Today it is being used for evaluations of dry-landing structural concepts for the return of the Orion spacecraft scheduled to journey to the moon by 2020. (Photo courtesy of NASA)

The East Area of NASA Langley, located at the Langley Air Force Base, was the birthplace of NACA laboratories in 1917. The photograph shows a mix of Air Force and NASA facilities. The huge Langley Full Scale Tunnel (white building near the waterfront in left foreground), Langley 20-Foot Spin Tunnel (cylindrical building next to the full-scale tunnel), and Langley 12-Foot Low-Speed Tunnel (spherical building next to the spin tunnel) will be discussed throughout this book. The NASA West Area is visible at the upper right. (Photo courtesy of NASA)

the nation's first civil aeronautical research laboratory in 1917, the facility was known as the Langley Memorial Aeronautical Laboratory under the management of the National Advisory Committee for Aeronautics (NACA), which had been formed by Congress in 1915. The membership of NACA was directed by the language of Congress: "The President is authorized to appoint not to exceed twelve members, to consist of two members from the War Department, from the office in charge of military aeronautics; two members from the Navy Department, from the office in charge of naval aeronautics; a representative each of the Smithsonian Institution, of the United States Weather Bureau, and of the United States Bureau of Standards; together with not more than five additional persons who shall be acquainted with the needs of aeronautical science, either civil or military, or skilled in aeronautical engineering or its allied sciences: Provided, That the members of the Advisory Committee for Aeronautics, as such, shall serve without compensation."

The charter of the NACA included the following excerpt:

"That it shall be the duty of the Advisory Committee for Aeronautics to supervise and direct the scientific study of the problems of flight, with a view to their practical solution, and to determine the problems which should be experimentally attacked, and to discuss their solution and their application to practical questions."

NACA quickly moved to staff its new facility and provide it with unique testing methods, laboratories, wind tunnels, and research aircraft. The scope of work initiated at Langley covered all the critical disciplines for aeronautics, including aerodynamics, structures and

Aerial view taken in 1950s shows the relative location of the NASA West Area (upper center) and the East Area (lower center). The Langley Full Scale Tunnel is at the bottom of the photograph. (Photo courtesy of NASA)

materials, propulsion systems, and flight controls. Although certain aspects of Langley's early responsibilities were transferred to other NACA spin-off centers during World War II, the recognized expertise of Langley's staff in a wide variety of aeronautical disciplines has remained a characteristic of the center to this day.

Langley researchers have traditionally conducted studies in both fundamental and applied aeronautics. Fundamental research in the discipline of aerodynamics involves the understanding, prediction, and control of the basic physics of fluid flows. The focus of applied aerodynamics involves studies of the aerodynamic characteristics of aircraft components and aircraft configurations. Perhaps the most important point for the reader to learn from this publication is that NASA researchers do not design aircraft as an objective of their research. Rather, they provide the design data, facilities, and methods that can be used by the aircraft industry in the design and development of aircraft. Results of experimental or computational studies of the relative performance of advanced airfoil shapes, for example, are summarized in technical reports and quickly transmitted to industry peers. Oral communications and face-to-face meetings are established to ensure the timely and effective dissemination of information. In many cases, Langley researchers are invited to participate in aircraft development programs as advisers and evaluators. This unique arrangement of mutual support between a government agency and U.S. industry has been an extremely productive mechanism for advancing the state of the art in aeronautics for both military and civil aviation.

Among the most important responsibilities of NACA and NASA have been the conception, construction, operation, and upgrading of critical aeronautical test facilities. This inventory includes items such as wind tunnels, computational facilities, structural test laboratories, piloted simulators, engine test stands, avionic and control-system laboratories, aircraft flight-test sites, and research aircraft. In the early days of aviation and during the World War II era, industry could not afford the cost and luxury of owning, operating, and maintaining such facilities, particularly for long-term or fundamental research projects. Thus, government facilities provided key services by reducing risk in the development of new aircraft designs, and supplying problem-solving efforts for operational civil and military aircraft. Requests from industry for entries into the government facilities for assistance in cooperative testing of concepts of mutual interest have usually been accepted on a fee-paying or cooperative basis. In addition, the Department of Defense (DoD) has usually been granted priority for projects associated with military programs. This historical relationship has been of great benefit to both parties involved and has been supported by all levels of management. In return for their contributed expertise, analysis, and facility-generated data, the researchers of NACA and NASA benefit from joint funding of aircraft models and other equipment, as well as obtaining invaluable feedback data from aircraft development programs regarding the accuracy and interpretation of their predictions. This cooperative scientific culture is the basis upon which most of the radical aircraft experiences discussed in this book occurred.

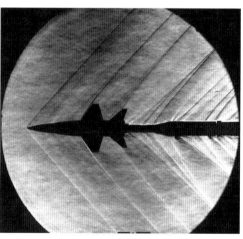

Langley conducts wind-tunnel aerodynamic research and maintains advanced wind tunnels to serve the Nation's requirements. The stable of tunnels covers all speed ranges from subsonic to hypersonic conditions. Here, shock waves created by flow over a model of the X-15 research aircraft in the Langley 4 x 4 Foot Supersonic Tunnel in 1962 are visualized using a schlieren optical technique. Note the shock waves generated from the sting mount as well as the model. The interactions and coalescing of shock waves influences the sonic-boom signature of supersonic aircraft, as will be discussed in Chapter 6. (Photo courtesy of NASA)

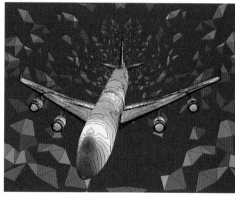

The emergence of powerful computer-based analysis methods developed by Langley and others has added unprecedented design-analysis capability for the aeronautical community. The use of Computational Fluid Dynamics (CFD) methods has reduced the number of wind-tunnel tests required for aircraft-development programs and allowed the designer to focus on configuration details and conditions not covered in tunnel tests. Also, the wind-tunnel tests can be directed at complex flow conditions not amenable to computational solutions at the present time. The two tools complement each other, and wind tunnels are not likely to go out of business anytime soon. This graphic depicts a computational solution for pressures acting on a transport configuration. The computer method, known as the award-winning NASA Tetrahedral Unstructured Software System (TetrSS), was developed by Langley and has been used extensively by hundreds of industry and government teams since 1991. Use of the tetrahedral structures depicted by the graphic to solve flow properties is a major computational breakthrough. (Graphic courtesy of NASA)

One of the most visible indications of aeronautical research at Langley has been its unique inventory of wind tunnels. In general, the public has little understanding of the capabilities and applications of wind tunnels or the need for so many specialized tunnels that have been put into operation over the years. Wind tunnels are designed to create test capabilities that permit researchers to measure, observe, and analyze aerodynamic behavior of advanced aircraft in specific speed ranges, including subsonic, transonic, supersonic, and hypersonic speeds. They are also designed in many cases to permit unique tests such as evaluations of aircraft spinning and spin-recovery characteristics, or assessments of aeroelasticity and airframe flutter. Langley's wind tunnels have covered these specialty areas and many more over the years. At one time in the early 1990s, the center operated more than 30 different types of wind tunnels. When Langley celebrated its 90th birthday in 2007, many tens of thousands of tests had been conducted in its current inventory of tunnels and the numerous facilities that preceded them.

Arguably, three of Langley's wind tunnels have had the most productive history pertinent to the testing and evaluation of radical aircraft configurations. The Langley Full Scale Tunnel, the Langley 20-Foot Vertical Spin Tunnel, and the Langley Free-Flight Tunnel have played key roles in the research results used to assess advanced configurations, and they are featured in this book. Because these historic tunnels will be mentioned so frequently in subsequent sections, a brief review of some of their features and test capabilities is now provided to help the reader visualize and interpret the testing scenario for the discussions to follow.

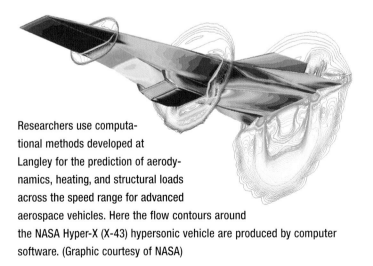

Researchers use computational methods developed at Langley for the prediction of aerodynamics, heating, and structural loads across the speed range for advanced aerospace vehicles. Here the flow contours around the NASA Hyper-X (X-43) hypersonic vehicle are produced by computer software. (Graphic courtesy of NASA)

The Langley Full Scale Tunnel

The Langley Full Scale Tunnel is an extraordinarily large subsonic wind tunnel that has been in operation for more than 77 years, producing invaluable information with emphasis on the performance, stability, and control of unconventional aircraft configurations. As will be revealed in later sections of this book, test programs conducted in this tunnel have investigated the subsonic aerodynamic characteristics of sub-scale models or actual full-scale versions of virtually every type of fixed-wing aircraft, from early biplanes to proposed configurations for future supersonic transports and advanced high-performance fighter aircraft.

As early aeronautical research efforts accelerated at Langley in the 1920s, NACA constructed pioneering wind tunnels that began to unlock the secrets to significant advances in aircraft technology. Heading the list was the Langley Variable Density Tunnel, which provided a revolutionary new capability in 1922 to evaluate aerodynamic performance using sub-scale models under pressurized conditions that resulted in more accurate simulation of a full-scale aircraft in atmospheric flight. In addition, the Langley 20-Foot Propeller Research Tunnel led the way for tunnel testing of full-scale aircraft

The Langley Full Scale Tunnel was put into operation by the NACA in 1931 as the largest wind tunnel in the United States. In continuous testing for over 77 years, the tunnel is still the site of aerodynamic research under the management of the Old Dominion University. The white office building and its adjacent gray wind tunnels to the left of the full-scale tunnel is the site of the 8-Foot High-Speed Tunnel where Richard T. Whitcomb conceived and tested his famous area rule; and later the 8-Foot Transonic Tunnel where he developed winglets and supercritical airfoils. (Photo courtesy of NASA)

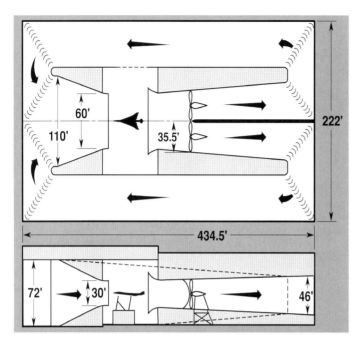

components, including engines and engine-cowl arrangements. However, scientific issues remained over the disagreements sometimes observed between sub-scale tests and flight results. Of special concern were the potential effects of aircraft geometric features, such as slots, gaps, protuberances, wires, and other features that were difficult or impossible to fabricate at model scale. The ensuing arguments of the validity of sub-scale model tests normally ended in a perceived need for a wind tunnel capable of testing full-scale aircraft of the time.

Cut-away top and side cross-sectional views of the Langley Full Scale Tunnel showing dimensions of the test section, location of the two drive fans, and the return paths of the closed-return tunnel. Note the open test section in which the air enters the entrance cone of the 30- by 60-ft test section and exits through the exit cone. The lack of walls immediately around the airstream permits many unique tests to be accomplished. (Graphic courtesy of NASA)

View of the attendees at the NACA Ninth Annual Aircraft Engineering Research Conference of May 1934 in the full-scale tunnel. A Boeing P-26 was mounted in the tunnel for tests during the period. This historic picture captures one of the most notable gatherings of famous aeronautical leaders ever taken. (Photo courtesy of NASA)

In the late 1920s, interest within NACA management for such a wind tunnel intensified, and a new tunnel was designed and constructed. Built with Depression-era funds and designed by Langley engineer Smith DeFrance, the gigantic Langley Full Scale Tunnel is a truly impressive sight, housed in a huge structure on the shores of the Back River in Hampton. The tunnel design is known as a closed-circuit, open-throat tunnel, characterized by an open quasi-elliptical test section with dimensions of 60 feet across by 30 feet high and a length of 56 feet. Air is drawn through the test section at subsonic speeds by two four-blade 35.5-foot-diameter propellers powered by two 4,000-horsepower electric motors. Designed and initially operated for test speeds of about 100 mph, the top speed for testing has been reduced over the years due to structural limitations to a current speed of about 80 mph. Once it passes through the test section, the moving airstream returns to the test section in a bifurcated manner through wall passageways within the building enclosing the test section.

The dedication ceremony for the opening of the wind tunnel occurred in May 1931, and the facility immediately became a showplace for NACA to demonstrate its impressive research facilities. NACA hosted annual "inspections" by the industry, military, and academic communities at Langley to brief the attendees on progress in aeronautical technology, and the group usually stopped at the full scale tunnel for lectures and observations of tests in progress. Attendees at the inspections were technical leaders, decision makers, and very influential individuals. Shown in the accompanying photograph are visitors to the full-scale tunnel during the NACA Ninth Annual Aircraft Engineering Research Conference in May 1934. This notable gathering included such legendary aviation figures as Orville Wright and Charles Lindberg (both members of the NACA committee), Howard Hughes, Leroy Grumman, John Northrop, Alexander de Seversky, Harold Pitcairn, Lloyd Stearman, Henry Berliner, Giuseppe Bellanca, Theodore Wright, Sherman Fairchild, James Doolittle, Elmer Sperry, Clarence Taylor, and Grover Loening. Also of historical interest is the fact that three of the original staff members of the full-scale tunnel in 1931 subsequently became directors

The Langley Full Scale Tunnel was the centerpiece of drag reduction studies for U.S. military aircraft during World War II. Over 30 different fighters, scout planes, and torpedo bombers were tested, resulting in recommended changes to significantly reduce drag and improve performance and top speed. This photo shows a P-51B Mustang in the sealed and faired condition awaiting drag-cleanup studies in September 1943. Note the tape used to seal all cracks and crevices on the surface panels. (Photo courtesy of NASA)

of other NACA or NASA centers later in their careers (Smith J. DeFrance-Ames, Abe Silverstein-Lewis, and Harry J. Goett-Goodard Space Flight Center).

Full-scale aircraft the size of World War II fighters could be accommodated on mounting struts, which were directly connected to scale-beam structures that permitted the measurement of air loads on the aircraft under test. Test procedures were developed to permit power-on testing of propeller-driven (and subsequently jet-driven) aircraft. During a typical test program, a series of special tests was conducted depending on the area of interest. For example, cruise-performance investigations involved measuring aerodynamic drag for the total airplane as well as the incremental drag effects from individual components, protuberances, and surface conditions. In addition, engine cooling and propeller performance at simulated cruise or during climb were studied using power-on tests. At the other end of the envelope, low-speed high-lift characteristics were measured to determine an aircraft's maximum lifting capability, stall speed, and the effects of high-lift

devices such as flaps and wing leading-edge slats. Tests were also conducted to evaluate longitudinal and lateral-directional stability and control. For stability tests, the contributions of aircraft components such as the horizontal tail were frequently determined by removing the tail and repeating the test program. In many cases, the direct measurements of air loads were augmented with extensive measurements of air pressures at critical locations on the vehicle. Armed with this information, project engineers could determine the nature and location of airflow conditions such as abrupt separation of smooth airflow over wings, or the onset of compressibility over a fuselage canopy.

The large size of the tunnel's test section and the lack of any wall structure immediately surrounding the open-jet airstream presented researchers with new capabilities and versatility for aerodynamic testing. In the mid-1950s, the tunnel was modified for free-flight model tests to evaluate dynamic stability and control characteristics using a unique testing technique that NACA had conceived and developed in other smaller Langley wind tunnels during the 1930s.

The free-flight technique has proven to be an excellent method of obtaining qualitative evaluations of the flying characteristics of aircraft configurations, especially unconventional and radical designs for which

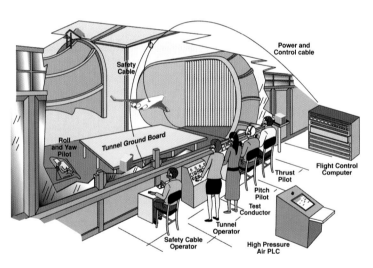

Sketch of setup for free-flight model testing in the Langley Full Scale Tunnel. The model flies without restraint in the open-throat test section while being powered by compressed air and control signals via a lightweight flexible cable. Piloting tasks are divided among test participants located in a balcony or in an enclosure in the tunnel exit cone. At the beginning and the end of the test, the model hangs from a steel safety cable. A digital flight-control computer is used to simulate control laws used by the full-scale airplane. The technique has proven to be especially valuable in assessing the flying characteristics of radical configurations. (Graphic courtesy of NASA)

Free-flight model of the YF-22 fighter undergoes stability and control flight tests for high-angle-of-attack conditions in 1994. Note the nose boom equipped with angle of attack and angle of sideslip sensors for control feedback. The flexible plastic tubing provides compressed air for simulation of jet engines. Control-signal wiring is taped to the tubing and the steel safety cable can be seen at the top of the picture. Lockheed Martin praised the results of these tests, which vividly demonstrated the high-angle-of-attack maneuverability of the YF-22. Results of subsequent flight tests of the full-scale airplane were in excellent agreement with the free-flight results. (Photo courtesy of NASA)

experience and data are lacking or nonexistent. In the free-flight technique, the powered model is flown without restraint in the 30-by 60-foot open-throat test section of the tunnel while being remotely controlled by human pilots. Three pilots are typically used during the tests because of the rapidity of the model motions (properly scaled models exhibit angular motions that are much faster than the full-scale aircraft). One pilot, who controls the model in roll and yaw, is located in an enclosure at the rear of the test section, while the pilot who controls the model in pitch is stationed in a balcony enclosure at one side of the tunnel along with the thrust pilot. Operators in the balcony also control the safety cable and actuate model variable-geometry features. Pneumatic and electric power and control signals are supplied to the model through a flexible trailing cable that was made up of wires and light plastic tubes. The free-flight technique also incorporates a steel cable that passes through a pulley above the test section. This element of the flight cable is used to restrain the model when an uncontrollable motion or mechanical failure occurs. The entire flight cable is kept slack during the flights by a safety-cable operator using a high-speed pneumatic winch.

In its 77-year history, the Langley Full Scale Tunnel has been the site for a continuous stream of aerodynamic investigations of aircraft configurations. From the biplane era of the 1930s and World War II military aircraft of the 1940s, research projects conducted in the tunnel have included Vertical and Short Takeoff and Landing (V/STOL) aircraft, parawing vehicles, advanced supersonic transport configurations, new military aircraft concepts, lifting bodies, and a large number of non-aerospace applications, including measurement of aerodynamic loads on buildings and drag-reduction studies for submarines. In recognition of the historic contributions and activities conducted in this facility, the United States Department of the Interior designated it a National Historic Landmark in 1985.

Faced with growing budget concerns in 1995, NASA retired the full-scale tunnel from its active inventory of wind tunnels and transferred operation of the facility to the Old Dominion University (ODU) of Norfolk, Virginia. Although aircraft testing in the tunnel has continued on a more limited basis to the present day, the focus of the ODU projects has been automotive and truck testing, including aerodynamic drag assessments and improvements for NASCAR racers.

The Langley 20-Foot Vertical Spin Tunnel

In the early days of manned flight, one of the most dreaded phenomena of aircraft operations was the tailspin. In the 1920s pilots had little background or experience with spins and the proper techniques to recover from them. Even more importantly, engineers had little understanding of the design features and physical factors that caused dangerous spins. The British were first to move out in this area, first using catapult-launched models to study spins, but they quickly found the technique to be time consuming with frequent repairs to models. In addition, the catapult launched-model technique was lacking because of the brief flight time for analysis before ground impact. Shortly thereafter, experiments began with a vertically oriented wind tunnel. In the United States, Charles H. Zimmerman, a brilliant NACA Langley researcher whose famous career as an aircraft designer will be discussed in a later chapter, became the principal advocate for spin research. At his initiation, Langley constructed a five-foot diameter "spin tunnel" in the late 1920s that was capable of measuring aerodynamic loads on aircraft models during spinning motions that were forced by a drive motor. Flow in this tunnel was vertically downward.

Zimmerman quickly followed with the design and construction of a 15-foot free-spinning tunnel in 1935 that permitted researchers to observe the spin and recovery motions of small balsa models. In contrast to the earlier five-foot tunnel, the airflow in the 15-foot tunnel was vertically upward to simulate the downward

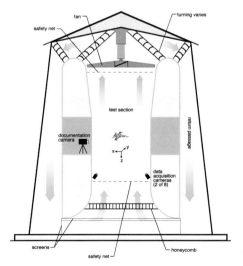

Cross-sectional view of the spin tunnel shows the closed-return tunnel configuration, the location of the drive fan at the top of the facility, and the locations of safety nets above and below the test section to restrain and retrieve models. The tunnel operator uses a rapid-response airspeed control to maintain the position of the model within the field of view of data-acquisition cameras.

velocity of an aircraft during spins. At the beginning of a typical test, the model was mounted on a launching spindle at the end of a wooden rod held by a tunnel technician. A second tunnel operator controlled the airspeed so that the air forces on the model equaled its weight, after which the model automatically disengaged from the spindle and continued to float as the airspeed was adjusted to maintain the model position at eye level in the test section. The model control surfaces were moved to pre-determined anti-spin settings by a clockwork mechanism, and the rapidity of recovery from the spin was noted. After the test was completed, the model was lowered into a large net at the bottom of the test section and recovered with a long-handled clamp.

These early spin-research facilities were replaced in 1941 by the Langley 20-foot Vertical Spin Tunnel, which has been the scene of continuous evaluation processes for new airplane designs. Nestled next to the enormous Langley Full Scale Tunnel building, the current facility features a vertically rising test airstream into which free-flying unpowered aircraft models are hand-launched to evaluate spinning and spin-recovery behavior, tumbling resistance, and recovery from other out-of-control situations. The tunnel is a closed-throat, annular return wind tunnel with a 12-sided test section 20-feet across by

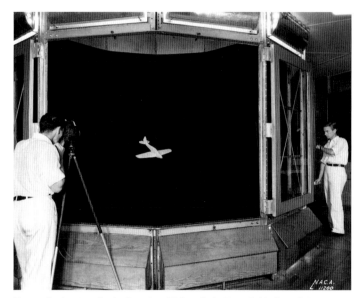

Free-spinning test in the Langley 15-foot Spin Tunnel. Motion-picture records were made after the model reached a steady spinning condition and after spin-recovery controls were actuated by a clockwork mechanism. Measurements were then made of the rate of spin, attitude of the model, and the number of turns required for recovery from the spin. (Photo courtesy of NASA)

The cylinder-shaped Langley 20-Foot Vertical Spin Tunnel and its associated office building located adjacent to the full-scale tunnel. The spin tunnel became operational in 1941 and has conducted spin and recovery studies for about 600 different aircraft designs. Also shown in the photograph is a sphere that encloses the Langley Free-Flight Tunnel (now enamed the Langley 12-Foot Low-Speed Tunnel). (Photo courtesy of NASA)

25-feet tall. The test-section air speed can be varied from zero to approximately 85 feet per second. Airflow in the test section is controlled by a three-blade fan powered by a 400-horsepower direct-current motor located in the top of the facility. The airspeed-control system is designed to permit rapid changes in fan speed to enable precise location of the model in the test section.

The models for free-spinning tests are launched with pre-rotation (similar to a Frisbee) into the vertically rising airstream, and a tunnel operator controls airspeed to stabilize the spinning model in view of observers. Images of the stabilized spinning motions are documented on high-resolution video calibrated to measure pertinent parameters including angle of attack, spin rate, and control positions. Spin-recovery characteristics of aircraft are studied by using remote actuation of the aerodynamic control surfaces of the models. Sizing of emergency spin-recovery parachute systems for flight-test aircraft is also determined by deploying various parachute configurations from the spinning model.

The 20-foot spin tunnel has supported the development of nearly all U.S. military fighter and attack aircraft, trainers and bombers during its 67-year history, and the total number of projects tested to date is about 600. Every radical aircraft design to be discussed in this

A model of the F-22 Raptor undergoes spin and recovery tests in 1996. The technician on the left has increased the speed of the tunnel so that the aerodynamic force is approximately equal to the weight of the model. The technician on the right has opened a door to the test section and has hand-launched the model with a spinning motion into the vertically rising airstream. The door is then closed, and the characteristics of the equilibrium spin (or spins) are noted. At a specified time, the technician in the background applies spin-recovery controls via a radio-control system. Scale emergency spin-recovery parachutes are also packed on the models and released during spins to provide recommendations for spin-test systems for the full-scale aircraft. Data generated in spin-tunnel tests forms the basis for flight-test plans in aircraft development programs. Over 600 different configurations have been tested to date. (Photo courtesy of NASA)

The Langley Free-Flight Tunnel

In the late 1930s, after successfully advocating for, designing, and initiating operations of the 15-foot spin tunnel, Charles H. Zimmerman conceived and successfully developed yet another unique wind-tunnel apparatus to study the dynamic stability and control characteristics of an aircraft model in a free-flying condition. A five-foot-diameter "proof-of-concept" wind tunnel was constructed and suspended from a yoke that permitted it to be rotated about a horizontal axis and tilted from the horizontal position to a nose-down orientation up to an angle of 25 degrees. A tunnel operator stood at the side of the test section and controlled the tilt angle of the tunnel as well as the airspeed produced by a fan located at the right rear of the test section. His main function was to adjust the airspeed and tunnel angles so that the model remained stationary in the center of the tunnel during a test. The evaluation pilot was positioned at the rear of the tunnel, where he could easily see the model's lateral motions and provide inputs to the model's controls via fine wires that were kept slack during the flight.

In a typical free-flight test, the model was placed at the center of a takeoff platform and the elevator or

book was tested in the spin tunnel. The tunnel has also been used for tests of planetary probes, lifting bodies, and spacecraft. Experience has shown that spinning and spin-recovery characteristics may be dramatically influenced by seemingly minor geometric variations in aircraft wings, tails, and other components; therefore, test projects lasting several months are common occurrences in the tunnel. Weight distribution and external stores may also have profound effects, and carefully designed test programs are conducted to determine and assess all spin modes for potentially unsatisfactory and dangerous behavior.

Unconventional aircraft designs, such as tailless flying wings or highly unstable configurations, may be susceptible to tumbling, a phenomenon in which the aircraft rotates in on-going, uncontrollable nose-over-tail gyrations. During World War II, tumbling was exhibited by several early flying-wing designs, and Langley researchers included the evaluation of tumbling as a standard operating procedure for testing these types of designs in the spin tunnel.

Zimmerman's "proof of concept" 5-foot free-flight tunnel during tests for a model of the Brewster XF2A-1 Buffalo fighter. The tunnel operator at the left controls the tilt angle of the test section and the airspeed of the tunnel while the evaluation pilot positioned at the rear of the tunnel provides control inputs and evaluates the model's dynamic stability and control characteristics. With highly successful results from this pilot tunnel, Zimmerman designed and developed a larger 12-foot free-flight tunnel. (Photo courtesy of NASA)

model pitch control was manually adjusted to a desired setting. The tunnel angle was adjusted to the expected glide path of the unpowered model, and the airspeed was slowly increased until the model rose from the platform and assumed a flying attitude. The tunnel operator and evaluation pilot coordinated their tasks to permit an assessment of the relative stability and responses of the model to control inputs.

Initial testing in the five-foot free-flight tunnel started in 1937 with very encouraging results, including the development of an automatic light-beam-control device to reduce crashes during the learning process. Zimmerman's team quickly designed and developed a larger

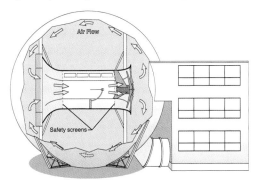

Cutaway drawing of the interior of the sphere enclosing the free-flight tunnel, showing the path of airflow in the closed-circuit tunnel and the location of the drive fan. Located adjacent to the engineering offices, the tunnel entrances provided easy access during test programs. (Photo courtesy of NASA)

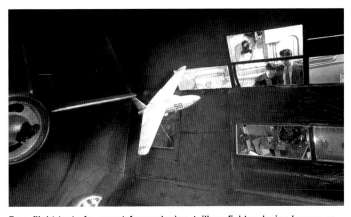

Free-flight test of a swept-forward wing, tailless fighter design known as the Air Materiel Command MCD-387-A configuration in the Langley 12-Foot Free-Flight Tunnel. Free-flight tests provide valuable information on the dynamic stability and control characteristics of radical aircraft configurations for which no experience or data are available. When the free-flight test technique was transferred to the Langley Full Scale Tunnel, the test section of the free-flight tunnel was fixed in a horizontal attitude and the tunnel has been subsequently used as a low-cost facility for conventional wind-tunnel tests to determine aerodynamic forces and moments. (Photo courtesy of NASA)

free-flight capability for a 12-foot tunnel with an octagonal test section in 1939. Housed in a sphere near today's spin tunnel, the current Langley 12-Foot Free-Flight Tunnel (now known as the Langley 12-Foot Low-Speed Tunnel) provides all-weather test capability adjacent to an office building staffed by researchers and support crews. The closed-circuit free-flight tunnel initially operated with a tiltable (hydraulically driven) test section in a two-operator mode similar to the earlier tunnel. For about 20 years, highly successful studies of radical aircraft configurations were conducted, during which advancements in internal strain-gauge technology permitted not only free-flight evaluations but also measurements of the aerodynamic characteristics of the free-flight models during conventional wind-tunnel tests.

In 1958, a slack in the post-war operational schedule of the Langley Full Scale Tunnel permitted exploratory free-flight tests to be conducted in its gigantic 30- by 60-foot test section, providing much more freedom and less risk during flight tests. In addition, the models could be much larger and sophisticated than the simple balsa models used in the free-flight tunnel. For these reasons, the free-flight technique was transferred to the larger wind tunnel. Dismayed by the arrival of the relatively small flying models, long-time technicians at the full-scale tunnel referred to the new invaders as "Butterflies." When the technique was transferred, the free-flight tunnel was converted for conventional force-test studies of aerodynamics and its test section was fixed in a horizontal attitude. The tunnel has continued to soldier on as an exploratory low-cost test facility for more than 50 years, and is in continuous use by NASA Langley today.

The full scale tunnel, vertical spin tunnel, and free-flight tunnel are only three of the many past and current historic wind tunnels of the NACA and NASA inventory at the Langley Research Center. Even a cursory discussion of the contributions and activities of the other Langley tunnels is far beyond the intended scope of this book. The reader can be assured that the other tunnels (and those at the other NASA centers that conduct aeronautics research) also have many more contributions and stories to be told.

With this background information in mind, let us recount how the NACA and NASA research staffs and their unique facilities participated in some of the more interesting developments of radical aircraft configurations over the past 70 years.

NO.75 10-27-36
BELL AIRCRAFT CORP.
PROJECT M-9-35

CHAPTER 1

THE EARLY DAYS

Following the successful debut of the airplane in World War I and the subsequent growth of aviation during the 1920s, aircraft designs advanced in complexity from fabric-covered biplanes to aluminum-covered monoplanes. Designs that were considered to be "radical" by ordinary standards of the period emerged in both the civil and military sectors.

Concepts that pushed the limits of flight, in terms of safety, speed, maneuverability, and range, appeared virtually every day. With this growth and advancement of aviation came new innovations, conceived by pioneering figures in the world of aeronautics. General aviation aircraft designs that made use of new revolutionary aerodynamic devices designed to improve performance and enhance safety emerged on the national scene.

On the military front, highly maneuverable radical streamlined fighter designs appeared, as well as mammoth bomber designs that maximized the range and reach of airpower. While many of these innovative designs represented significant steps forward in aviation, several failed to make it into widespread production. At the NACA Langley Memorial Aeronautical Laboratory, fundamental aerodynamic research was performed on many of these designs, and researchers were busy providing data on the effects of wing shape, airfoils, and other design variables for the use of industry designers. By the early 1930s, the NACA had acquired expertise in several areas of aeronautical research. Researchers were performing routine studies of the spin characteristics of aircraft designs in the laboratory's first spin tunnel, and hydrodynamic tests of seaplane and flying-boat hulls,

pontoons, and floats were underway in a large towing basin. Many radical aircraft designs underwent aerodynamic testing in the full-scale tunnel, supplemented by extensive flight investigations conducted by the Langley Flight Research Division.

Lockheed Y1C-23

During the early 1930s, Lockheed pursued the development of a civil aircraft for use by Charles Lindbergh in surveying oceanic flight routes. Lindbergh requested

The Lockheed Y1C-23 at Langley for tests in the Langley Full Scale Tunnel in 1932. The aircraft, which could carry one passenger in addition to the pilot, had been bought by the Army Air Corps and used as an executive transport for high-level military commanders. Note the flag of the Assistant Secretary of War on the fuselage. The Y1C-23 featured an all-metal fuselage and retractable landing gear. (Photo courtesy of NASA)

that two different types of airplanes be considered: one with fixed undercarriage and the other with retractable undercarriage. The primary aircraft developed for Lindbergh was the Lockheed Sirius, which had fixed-landing gear. The Sirius became the base design for the Altair, which was one of the first aircraft in the world to use retractable landing gear. Lindbergh eventually rescinded his request for the aircraft with retractable undercarriage, and the aircraft was procured by the Army Air Corps and given the designation Y1C-23. The single Y1C-23 differed from other versions of the Altair in that it had a metal fuselage (other previous versions had wooden fuselages). Based at Bolling Field at Washington, D.C., the aircraft was used to transport high-level military dignitaries. When tested at Langley, it carried the flag insignia of the Assistant Secretary of War. Although not a truly radical configuration, the NACA effort on the Y1C-23 is included to give an example of the type of fundamental research being conducted at Langley in the early 1930s.

In 1932, the Army requested that the NACA conduct tests of the Y1C-23 in the Langley Full Scale Tunnel. These tests were specifically designed to study the effects of retractable landing-gear openings in the underside of a wing on takeoff performance. Both Lockheed and the Army Air Corps had predicted that the openings were

The Lockheed Y1C-23 mounted in the Langley Full Scale Tunnel for studies of the potential performance penalties caused by the wheel wells of the retractable landing gear. The results showed that the drag of the open wheel wells was not as large as expected.

significantly degrading aerodynamic performance. In fact, the aircraft had gained notoriety as requiring an excessive takeoff distance. The airplane's landing gear was completely housed in the wing when retracted, but when it was extended the resulting openings in the lower wing had an area equal to the side area of the struts and wheels. During the full-scale tunnel tests, lift and drag characteristics were evaluated at various angles of attack with landing gear up and down. Performance assessments were made at an air speed of 60 miles per hour with the landing gear up, landing gear extended and wheel wells open, and with landing gear extended and the wheel wells covered with sheet metal.

During the brief two-week full-scale tunnel test program it was found that concealing the wheel openings with sheet metal while the landing gear was down reduced drag by only 2 percent at take-off conditions. As a result, Langley researchers concluded that the wheel openings had little or no impact on the aircraft's performance during take-off. They suggested that the landing gear's design could be aerodynamically refined to reduce drag; however, making such a modification might adversely affect the landing characteristics of the aircraft. Furthermore, consideration as to the incorporation of some other mechanism in the aircraft's design might be required to shorten the necessary landing distance. However, a more significant finding was that with the landing gear in the retracted position, the minimum drag of the aircraft was reduced by 50 percent. This result sparked considerable research on drag reduction for World War II aircraft subsequently tested in the tunnel.

The McDonnell Doodlebug

The year 1929 represented a high point in aviation. New monoplane aircraft designs were appearing across the globe, and revolutionary innovations such as the NACA cowling that had been conceived and developed by a team of engineers led by Fred Weick at Langley were beginning to be applied to aircraft designs. In response to the rapid growth of aviation, philanthropists, such as Daniel Guggenheim, sponsored aviation competitions designed to boost the popularity and state of the art. In the winter of 1929, Guggenheim sponsored the Daniel Guggenheim International Safe Aircraft Competition, aimed at advancing safety by improving the art of airplane design and fabrication. The competition featured

Front view of the Doodlebug showing the aircraft's innovative safety features including wing full-span leading-edge slots, slotted trailing-edge flaps, and wide-span landing gear. Results of NACA tests in the full-scale tunnel identified fixes for tail buffet. (Photo courtesy of NASA)

numerous domestic and international designs, and the winner took home a prize of $100,000. In the 1929 competition, the field of 10 competitors included: the Curtiss Tanager biplane, which ultimately finished first; a Handley-Page aircraft, which finished second; and a unique fabric-covered monoplane design conceived by famous aviation pioneer and McDonnell Corp. founder James S. McDonnell.

McDonnell's two-place monoplane, known as the Doodlebug, incorporated several revolutionary features, including an automatic leading-edge slot across the entire span of the wing and a slotted trailing-edge flap across 70 percent of the wing, a wide-span landing gear for safer landings, and streamlined wing-support struts to enhance structural integrity with minimal drag. The trailing-edge flap and an adjustable stabilizer were interconnected by a single lever to minimize trim changes when the flaps were lowered or raised. The airplane also

The McDonnell Doodlebug at Langley after being procured by the NACA in 1931. The airplane no longer had its original NACA cowling. It also lacked wing fillets and therefore suffered from degraded performance and severe tail buffet. The NACA replaced the original internal aluminum fuselage structure due to excessive corrosion. Note the NACA emblem on the tail. (Photo courtesy of NASA)

used an NACA cowling to reduce drag and promote improved aerodynamic performance. The aircraft had a top speed of 110 mph and a stall speed of 35 mph with a landing ground roll of only 150 feet. While the design held much promise for safer flight operations, it experienced a series of operational setbacks (crashes during trial flights) that ultimately doomed the entry in the competition. However, in the few successful flights that McDonnell flew in the Doodlebug, the airplane demonstrated superior short-takeoff-and-landing characteristics and maneuverability compared to the other entries in the competition. Following the unsuccessful Guggenheim competition bid, McDonnell flew the aircraft in demonstrations for about a year and attempted to market the Doodlebug to commercial buyers, but this effort generated no "takers." The onset of the Great Depression also hindered any last hopes of being able to find a financial backer for the airplane. The NACA, however, took a keen interest in the technologies embodied in the Doodlebug for improved aviation safety, and decided to procure it for research at Langley in 1931 with the aircraft designation NACA 42.

The airplane, which had been dinged up considerably in publicity flights, underwent several modifications at NACA Langley. Included in these modifications were replacement of the corrosion-depleted original aluminum fuselage bracing, addition of wing fillets, and a new NACA cowling to replace the original cowling that had been destroyed. The wing fillets and new cowling were never flown on the aircraft, and were only installed for wind-tunnel tests.

NACA flight-test evaluations of the Doodlebug at Langley in 1931 revealed that although the performance of the slots and flap were impressive (total increase in lift of 94 percent), the airplane experienced severe tail buffeting. The Doodlebug was one of the earliest monoplanes, and worldwide concern had begun to arise over the cause and cures for adverse buffeting effects experienced by monoplane designs at the time. An earlier accident in Europe on another aircraft that disintegrated in flight had caused speculation that a very fundamental flow problem was occurring. After the Langley flight tests, a detailed test program on wing-fuselage interference was scheduled in the Langley Full Scale Tunnel, which was then into its second year of operations.

The Doodlebug wind-tunnel test program ran from December 1932 to February 1933, and the results became a classic design guide on the benefits of wing-fuselage fillets and other devices on aerodynamic performance and buffet alleviation. During the study an assortment of approaches to alleviate the wing-fuselage juncture flow problem were identified, including two different wing-fuselage fillets, an NACA cowling, a reflexed wing trailing edge adjacent to the fuselage, auxiliary airfoils of short span in three different positions ahead of the wing leading edge in close to the fuselage, and various combinations of the devices. The testing included power-on and power-off conditions during which the NACA test personnel sat in the cockpit holding a stick with a trailing wool tuft to observe the behavior of airflow at the wing root and tail of the aircraft. For the tunnel tests, the leading-edge slots and trailing-edge flaps were set to the neutral position and covered over.

Wing-fuselage fillets were designed by the Langley staff to control the rate at which airflow in the critical wing-fuselage juncture diverged and thereby prevent flow separation. As previously stated, the airplane was outfitted with a NACA cowling specifically designed for the tunnel tests. The reflexed trailing edge was created by extending the upward curvature of the lower surface of the wing and extending the upper surface to points of tangency with the modified lines. Vertical movements

The McDonnell Doodlebug undergoing tests in the Langley Full Scale Tunnel in 1931. During the tests, a new Langley-designed NACA cowling was applied, as well as wing fillets developed by the Langley staff. The modifications completely eliminated a severe flow-separation problem at the wing-fuselage juncture. This test program proved to be the textbook example for avoiding similar problems on monoplane designs. (Photo courtesy of NASA)

Rear view of the McDonnell Doodlebug with the NACA designation number NACA-42 visible on the upper right wing. Note the lack of aerodynamic fillets at the wing-fuselage juncture and the wing leading-edge slot. (Photo courtesy of NASA)

of the tip of the horizontal tail were recorded by instrumentation to measure the intensity of the buffeting, and the direction and speed of the airflow of the tail were also measured.

The airflow separation problem for the basic aircraft was readily apparent during flow-visualization studies, which showed that the wing-fuselage interference caused a premature stalling of the wing at the root. The problem was caused by several factors, including the tapering of the fuselage toward the rear and bottom, the additional friction drag caused by the fuselage, and the presence of the large, exposed engine cylinders. The reflexed trailing-edge concept had only a minor effect on the flow separation, and actually caused the tail buffet to increase. The auxiliary airfoils gave some improvement, but were inferior to the combined fillets and NACA cowling. The effectiveness of the fillets and the NACA cowling in preventing flow breakdown was quite apparent both in flow visualization studies and in overall forces and moments measured on the airplane. The tail buffeting levels were significantly reduced to amplitudes small enough to be considered unobjectionable throughout the range of angle of attack tested.

The results of this very fundamental flight and wind-tunnel investigation by the NACA at Langley have

since served as a textbook example of the causes and alleviation of wing-fuselage interference effects for monoplanes. The data generated by the study were quickly absorbed by the aviation industry and resulted in dramatic improvements in the performance and safety of U.S. aircraft in World War II and the general aviation fleet following the war.

Weick W-1 and W-1A

During the early 1930s, a team of engineers at NACA Langley, led by aviation pioneer Fred E. Weick, developed a radical homebuilt general aviation aircraft known as the Weick W-1. Weick had already become a legend for his pioneering contributions to the development of the NACA low-drag cowling that later became a standard feature on aircraft worldwide and that won the coveted Collier Trophy. He was deeply devoted to establishing the practicality of personal-owner aircraft, but recognized that safety was the paramount factor that would influence the acceptance of such aircraft. The idea for the W-1 design was born from meetings Weick held with his fellow colleagues who worked in the Langley 7- by 10-foot Wind Tunnel. During these meetings, ideas and concepts for the "ideal" general aviation aircraft were discussed. Specifications for performance, stability and control, and noise were established, and simplicity of design and operation were emphasized. Subsequently, the participants built free-flying models of several candidate concepts and test flew them in one of the cavernous return passages of the Langley Full Scale Tunnel in 1931. The group agreed that the aircraft concept that demonstrated the best performance and stability would be selected for full-scale production and eventual flight testing.

A pusher-propeller design, known as the W-1, won the competition. Equipped with a fixed auxiliary airfoil ahead of the leading edge of the wing and no trailing-edge flap, the two-place configuration exhibited a 30-percent increase in maximum lift and a 50-percent increase in maximum usable angle of attack compared to designs that used conventional wings. Weick was particularly concerned about the dangers of ground loops that were being experienced by contemporary aircraft during landings, and his NACA team devoted a substantial effort to the landing gear's design. After considerable study, Weick arrived at a configuration he named a "tricycle landing gear" in one of the most

Close-up views of the wing-fuselage fillets added to the Doodlebug design to eliminate flow separation. Langley wind-tunnel and flight-research studies of the Doodlebug resulted in design practices that enhanced future generations of monoplane configurations. (Photo courtesy of NASA)

noteworthy contributions in aviation history. The breakthrough in this landing-gear arrangement was a steerable nose wheel, which was a radical departure from other three-wheeled airplane designs of the time.

The W-1 featured a highly unconventional twin-boom pusher configuration that was covered in fabric. The flying gross weight (pilot plus oil and gas) of the airplane was 1,003 pounds. It made use of a Pobjoy radial engine and had two seats in tandem in the cockpit nacelle. Weick and his team began assembly of the W-1 in his garage during weekend work sessions, finally finishing the aircraft in late 1933.

The Bureau of Air Commerce had initiated a major effort to expand personal-owner flight operations in the United States during the early 1930s, and they were briefed on the W-1 design by Weick during a fact-finding mission to Langley. Impressed by the potential benefits of the configuration, the bureau requested that the NACA conduct wind-tunnel and flight tests to evaluate the characteristics of the W-1. The W-1 was quickly prepared for aerodynamic testing in the Langley Full Scale Tunnel, and the brief two-week test program was conducted beginning on March 1, 1934. The two-seat aircraft was mounted in the huge wind tunnel and its aerodynamic characteristics investigated without the propeller, and with rudders and ailerons set to neutral. Lift, drag, and stability were studied, with the major conclusion being that the only negative finding was low directional stability.

After the tunnel test, NACA test pilots conducted a brief flight evaluation of the W-1 in April, finding that it was very resistant to ground looping but confirmed that the directional stability was very low (as predicted by the tunnel tests). In addition, adverse yaw from the ailerons resulted in unsatisfactory roll control. Consequently, Weick and his team made modifications to the design, including increasing the size of the vertical tails.

One of Weick's ultimate design objectives was to devise a simple two-control concept for the W-1 to simplify coordination of the lateral-directional controls for inexperienced private pilots during turns. During an analysis of motions produced by various rudder and aileron connection schemes, he was assisted by a young Langley mathematician named Robert T. Jones, who would later gain fame for his pioneering work on the benefits of swept wings and the oblique (" switchblade") wing concept.

By 1935, a more advanced version of the W-1, known as the W-1A, was completed, and, in response to a

Close-up of the Pobjoy radial engine on the rear of the Weick W-1. Fred Weick was one of the most famous NACA researchers at Langley, and his interest and contributions to the general aviation community range from the first concepts for agricultural aviation to the Ercoupe design. (Photo courtesy of NASA)

In March 1934, the NACA conducted performance, stability, and control tests of the Weick W-1 in the Langley Full Scale Tunnel. The aircraft featured a highly unusual twin-boom pusher design, an auxiliary airfoil in front of the wing leading edge, and the first steerable tricycle landing-gear configuration. All results were positive, with the exception that directional stability was found to be low. (Photo courtesy of NASA)

Another view of the Weick W-1 in the test section of the full-scale tunnel. The full-scale tunnel tests led Weick to make modifications to the design, which significantly improved stability and control for his second-generation aircraft. (Photo courtesy of NASA)

formal request by the Bureau of Air Commerce, it was tested in the full-scale tunnel and in flight at Langley. Weick sold the W-1A aircraft to the Bureau of Air Commerce. The W-1A design abandoned the fixed auxiliary airfoil of the W-1, and instead included glide-control flaps that could be deflected to a maximum of 87 degrees in an effort to enable safer and more precise landings. The flaps extended from the tail booms outward to the wingtips. Lateral control was provided by unique "slot-lip" ailerons, which were spoilers with a slot that permitted the air to flow through the wing to the undersurface. This particular configuration was based on general research that Weick and his associates had conducted at Langley on lateral-control concepts. In the full-scale tunnel tests, which started on May 5, 1935, drag-evaluation tests were conducted with the slot-lip aileron in both open and closed configurations.

The Weick W-1A as it was flown in flight research studies at Langley in 1935. The aircraft featured a glide-control flap on the wing trailing edge and lacked the auxiliary airfoil found on the W-1. Also seen on the wing undersurface are the slots of the slot-lip aileron roll-control devices. Aircraft was photographed on the NACA hangar apron. Note the Langley Full Scale Tunnel in the background behind the engineer. (Photo courtesy of NASA)

To maintain the closed configuration, fabric was applied to conceal the openings. Forces on the flap control stick and elevator effectiveness were also measured. In the tests, researchers found that the slot-lip aileron decreased lift and increased drag, as expected. The hinge location of the flap was found to be too forward (20-percent chord), resulting in high stick forces, and directional stability of the aircraft was judged to be low but adequate. In addition, it was determined that balance of the aircraft would be hard to maintain during flight for the planned center-of-gravity location due to large diving moments produced when the flap was deflected.

During the test period, the aircraft was on display in the test section of the full-scale tunnel for the annual NACA aircraft conference. Testing was completed on May 25 and was followed by flight evaluations by the NACA pilots.

Although the W-1 and W-1A were one-of-a-kind aircraft and never made it into production, the designs were clearly ahead of their time. The steerable tricycle landing-gear arrangement was later used on military aircraft such as the Lockheed P-38 Lightning and Bell P-39 Airacobra, and on Weick's famous Ercoupe general aviation aircraft. Today, it is a common feature found on virtually all military, civil transport, and general aviation aircraft.

Boeing XB-15

Tremendous strides were also made in bomber aircraft during the 1930s. Designers sought to push the state of the art's limits by creating huge advanced

The Weick W-1A during tests in the full-scale tunnel. The test results confirmed the benefits of general research studies conducted by Weick and his associates at Langley on roll-control devices. (Photo courtesy of NASA)

Close-up of the tail section of the large XBLR-1 model tested in the Full Scale Tunnel in late 1935 and early 1936. The model was used by the NACA to provide specific data in support of the airplane development program, but it also served as a workhorse for numerous general research studies on engine-wing integration for improved performance and cooling. (Photo courtesy of NASA)

Dramatic photograph of the XBLR-1 model undergoing power-on aerodynamic tests in the full-scale tunnel. Aerodynamic performance was measured during simulated cruise and climb conditions. (Photo courtesy of NASA)

View of the huge XBLR-1 model with in-line engines mounted in the full-scale tunnel for aerodynamic testing in 1936. Full-scale tunnel tests of this configuration provided data for low-drag engine nacelle installations for multi-engine bombers. (Photo courtesy of NASA)

Yet another general research study with the XBLR-1 model modified with twin radial low-drag engine nacelle installations, mounted in the full-scale tunnel. Data were obtained on the drag and propulsive characteristics of large twin-engine airplanes. Results of these investigations were useful for development of design methods. (Photo courtesy of NASA)

General research study with the XBLR-1 model to evaluate characteristics of pusher-prop arrangements for large aircraft. Note the extended trailing edge of the wing and the attempt to fair the wing trailing edge to the nacelles. (Photo courtesy of NASA)

configurations capable of vast operational ranges. The largest of these designs emerged from a 1934 Army design feasibility study that sought bomber candidates with a maximum range of 5,000 miles. The aircraft, built by Boeing and designated the XBLR-1 (Experimental Bomber, Long Range), was the largest aircraft in America at the time, having a fuselage length of 87 feet, 7 inches and a wingspan of 149 feet. Later redesignated the XB-15 in mid-1936, the configuration featured enormous wings capable of carrying large amounts of fuel. The interior wing structure was large enough to permit repair crews access to the engines during flight. The airplane also featured several modern amenities for the crew, including heated crew compartments complete with beds, a cooking area, and a bathroom.

At Langley, a large model of the XBLR-1 was tested in the full-scale tunnel in late 1935 and early 1936 at the request of the Bureau of Aeronautics. These tests were focused on finding efficient low-drag installations for radial air-cooled engines on large bombers. At the time of the tests, the in-line engine had already demonstrated its aerodynamic efficiency for several aircraft designs, and it became imperative that the same kind of aerodynamic efficiency be pursued for radial-engine installations. During the investigation, conventional radial-engine nacelle installations were tested on the large model and compared to tests with new low-drag engine nacelle installations. The results of the investigation showed that with new low-drag engine nacelle installations, drag in the XBLR-1 design could be significantly reduced and the maximum lift-to-drag ratio enhanced. Researchers at Langley estimated that this reduction in drag would result in a five-percent increase in top speed. The low-drag installation involved placing the engines partially submerged in the wing with streamline fairings over protruding cylinder heads. The tests also investigated the efficiency of changing from a tractor-propeller configuration to a pusher-propeller configuration with the propellers at the same span-wise location, but located on the trailing edge of the wing. The XBLR-1 model was also used as a workhorse research tool for NACA general aerodynamic research during several testing entries in the full-scale tunnel. For example, in 1940 it was used for studies of drag and propulsive characteristics of twin-engine airplanes.

By 1938, the XBLR-1 had been designated the XB-15, and flight research studies of the aircraft were conducted by the NACA at Langley. The flight tests yielded valuable design information on gust loads and their effects on the structural components of large aluminum aircraft. In addition, flying-quality information for the XB-15 was generated and used with data for 15 other aircraft by Robert R. Gilruth in the formulation of a report entitled *Requirements for Satisfactory Handling Qualities of Airplanes*, which was recognized as the authoritative source for the topic during the World War II years.

Although the XB-15 was underpowered, it performed remarkably well. It was used to transport vital medical goods from Langley Field, Virginia, to the earthquake-stricken country of Chile on February 14-15, 1939, on a nonstop humanitarian flight. The aircraft later established a load-to-altitude record on July 30, 1939, when it successfully transported a payload of 31,205 pounds to 8,200 feet. Only one XB-15 was built, and it never saw action during World War II. It was, however, later used as a cargo transport during the war and was redesignated as the XC-105. The XC-105 was relegated to the scrap heap in 1945.

The Boeing XB-15 heavy bomber during flight-research studies at Langley in 1938. Flight research studies of the aircraft by the NACA yielded valuable information regarding gust loads and their effects on the structural components of large aluminum aircraft. Data on flying qualities was also gathered for use in developing handling-quality requirements. (Photo courtesy of NASA)

Large 0.50-scale Bell XFM-1 Airacuda-powered model used in aerodynamic tests in the full-scale tunnel. Note the lack of aerodynamic fairing at the nacelle-wing juncture. (Photo courtesy of NASA)

Bell XFM-1

During the mid-1930s, military planners became concerned about potential enemy long-range bombers attacking the U.S. mainland or other strategic locations such as the Panama Canal. Within the military, some advocated the need for a multi-seat, long-range "bomber destroyer" to combat this perceived threat. After receiving proposals from the Lockheed Aircraft Co. and the Bell Aircraft Corp., the Army awarded a contract to Bell for its innovative concept by Bell chief designer Robert Woods, designated the XFM-1 Airacuda. Woods had been a NACA researcher at Langley for a year in 1928, and he became a legendary designer at Bell,

The XFM-1 powered model mounted in the full-scale tunnel for aerodynamic testing in late 1936. The full-scale XFM-1's armament featured two 37mm cannon mounted in the forward portion of the wing nacelles for "bomber destroyer" missions. (Photo courtesy of NASA)

leading the company's efforts in historical military aircraft programs like the P-39 Airacobra and the X-1 supersonic research aircraft.

The radical XFM-1 twin-engine pusher aircraft design was a huge aircraft, with a wingspan larger than many contemporary bombers of the day, such as the B-25. Two pilots and three gunners manned this advanced design. Two of the gunners were located in the forward portion of the wing nacelles, and each aimed a 37mm cannon. The third gunner/radioman used a pair of 0.50-caliber machine guns located in blisters on the side of the rear fuselage. The XFM-1 was equipped with experimental turbosupercharged Allison engines. A total of 12 test aircraft were ultimately built and participated in flight tests conducted by the Army. Three of the aircraft were outfitted with tricycle landing gear.

At the request of the Army Air Corps, a large one-half-scale powered model of the XFM-1 underwent aerodynamic testing in the Langley Full Scale Tunnel from December 1936 to May 1937 to determine the aerodynamic characteristics of the design and the relative effectiveness of various cooling-system designs for the Prestone radiators and oil coolers. Design features of the aircraft, including the mutual interference effects of the wing/nacelle junctures, flaps, and landing gear were evaluated. In addition, the performance of aileron-control surfaces and aerodynamic drag associated with the supercharger air-cooler design were analyzed. Tests were conducted for both power-on and power-off conditions, and flow visualization observations were made with wool tufts. As would be expected, Robert Woods

Close-up view of the starboard wing of the XFM-1 model showing cooling inlet, nacelle shape, and landing gear. In addition to aerodynamic performance testing, the XFM-1 model was tested for relative effectiveness of various cooling-system designs for the radiators. (Photo courtesy of NASA)

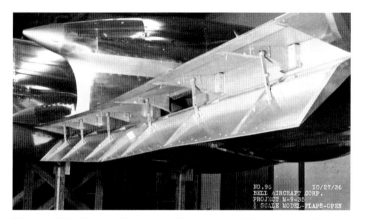

Closeup of the details of the wing trailing-edge split-flap system of the XFM-1 model. Tests included high-lift performance for various flap settings. (Photo courtesy of NASA)

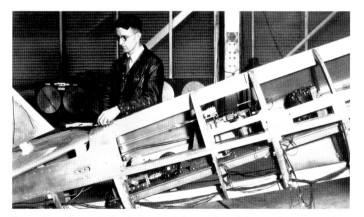

An engineer inspects the instrumentation and model control systems of the XFM-1-powered model. Extensive instrumentation was used to measure aerodynamic surface pressures, loads, and buffet as a common practice for tests in the full-scale tunnel. (Photo courtesy of NASA)

took a deep personal interest in the Langley test and visited on several occasions to provide concepts for cooling and aerodynamic improvements.

Langley researchers quickly discovered that the aerodynamic interference effects at the wing/nacelle junctures resulted in flow separation that seriously degraded aerodynamic performance, particularly during simulated climb conditions. This undesirable phenomenon was not corrected by propeller slipstream flow at high power settings. The flow separation near the wing's center section also resulted in unacceptable tail buffeting.

Wind-tunnel tests by the Army at Wright Field the previous year had indicated an appreciable improvement in climb performance when fillets were used at the wing/nacelle junctions; therefore, fillets were also evaluated in the full-scale tunnel tests with similar beneficial results. Fillets were subsequently incorporated into the full-scale airplane. Aerodynamic drag of the Airacuda model was assessed, along with an analysis of airflow through the supercharger air coolers, over the landing-gear door covers (the gear was in the retracted position), and over a new supercharger fairing. The results of the cooling-concept testing provided Bell with several options for cooling inlets and ducts either on top of the wing nacelles or in the leading edge of the wing. During the XFM-1 development, various arrangements were evaluated with guidance from the Langley tests.

The scope of power-on tests also included an evaluation of the effects of unsymmetrical power conditions as would be experienced following engine failure. The results of these tests did not reveal unexpected interference effects. The magnitude of the out-of-trim yawing moment measured in the tunnel agreed well with simple calculations based on engine thrust and the moment arm distance from the engine to the center of gravity of the airplane.

When the Airacuda made its maiden flight in September 1937, it sparked the public's interest due to its radical configuration and massive size. The event provided Bell with some badly needed publicity in its role as an upstart aviation company in a very competitive industry. Unfortunately, the XFM-1 had a large number of complex systems and operational issues. For example, the problem of crew escape during an emergency coupled with the aircraft's pusher propellers was a major concern. The gunners in the engine nacelles could leave their positions and enter the fuselage through a crawl way in the inner wing section to bail out from the fuselage door. In addition, the top speed of the XFM-1 was disappointingly low, and the cost of each aircraft was exceedingly high compared to other fighters of the day. More importantly, the perceived long-range enemy bomber threat from Europe or the Far East never became reality, and the mission for the XFM-1 vanished. For these reasons, the military lost interest in the aircraft and it never saw production. However, the widespread interest in its futuristic, radical design features provided recognition and financial stability to the Bell Corp. during the early days of heated industrial competition in World War II.

Grumman XF5F-1 Skyrocket

Along with the Army's interest in twin-engine fighters such as the XP-38, the Navy also initiated studies to explore the capabilities of twin-engine designs. The Bureau of Aeronautics requested Grumman to design a carrier-borne, twin-engine, fleet-defense fighter, resulting in the XF5F-1 Skyrocket. The aircraft's layout was remarkably radical for a Navy that was just beginning to transition from biplanes to single-engine monoplanes. A contract for Grumman to produce a single prototype was awarded in 1938.

In December 1938, a full-scale model of the Sky-rocket underwent a three-month series of detailed aerodynamic tests in the Langley Full Scale Tunnel at the request of the Bureau of Aeronautics. The objective of the test was to evaluate the effect of configuration variables, especially nacelle shape and location, on aerodynamic performance and stability. During these tests, the model was configured with its original engine nacelles placed at the original ("low") position as well as a new "high" position. In the high position, the nacelle was raised about seven inches so that the engine thrust line and wing chord were coincident. Tests of the model were also performed with modified shorter nacelles at a low position. The model's lift, drag, and pitching moments were also studied, with special attention being focused on various flap configurations, propulsion effects, and the model's behavior with counter-rotating

Full-scale model of the Grumman XF5F-1 Skyrocket undergoing aerodynamic tests in the full-scale tunnel in 1938. Nacelles are the original "long" configuration and vertical location. The nacelles were later shortened for initial flight tests. (Photo courtesy of NASA)

Nice front view of the XF5F-1 model during the test program. Note the simulated four-gun installation in the model's nose. The XF5F-1 program was eventually cancelled in favor of the development of the XF7F-1 Tigercat. (Photo courtesy of NASA)

Three-quarter rear view of the XF5F-1 model with landing gear extended. Trade studies on high-lift capability with long and short nacelles were conducted. Advantage of shorter nacelles was the capability to use full-span trailing-edge flaps for augmented lift. (Photo courtesy of NASA)

propellers. Various fuselage nose-shape modifications were studied as well. In addition, tail-off configurations were investigated for some of the tests.

The results of the study indicated that the most favorable nacelle location relative to high-speed performance was the high position, and the shortened nacelles provided the most favorable low-speed, high-lift characteristics. The latter result was directly attributable to the fact that the shortened nacelle permitted the flaps to be continuous, whereas the long nacelles required aerodynamically degrading structural cut-outs in the flap. The original XF5F-1 configuration featured much longer nacelles, but the first version of the airplane adopted the shorter nacelle recommended by the NACA tests. Researchers also found that power was destabilizing on longitudinal stability, and that the direction of propeller rotation or counter-rotating propellers had little effect on stability. Somewhat surprisingly, variations in the nose shape of the model did not result in significant effects on drag.

The XF5F-1's first flight occurred on April 1, 1940, and its impressive climb performance quickly earned it the nickname "Skyrocket." However, some difficulties were immediately experienced, the most serious involving engine cooling. Consequently, the aircraft's oil-cooling ducts were reconfigured. Other changes, such as a canopy height adjustment, new armament configuration, reconfigured engine nacelles (inspired by the suggestions offered by NACA Langley), and an extension of the fuselage ahead of the wing were also

made in the design. By 1942, additional flight tests revealed more problems, and Grumman became more interested in developing its new XF7F-1 Tigercat design, which represented a revolutionary step above the Skyrocket. The Tigercat was twice as powerful and twice as heavy as the XF5F-1. Nevertheless, the Skyrocket flew on in a number of additional tests involving more than 200 flights before eventually suffering a belly landing, forced by a landing-gear failure, on December 11, 1944. The XF5F-1 was withdrawn from service and the program was subsequently cancelled with only one prototype aircraft being produced.

Seversky XP-41

During the 1930s, new all-metal monoplane fighters replaced wood and fabric-covered pursuit biplanes. A radical new fighter design known as the XP-41 by the Seversky Corp. (which later became the Republic Aviation Corp.) appeared in 1938. The aircraft was similar to the design of the company's P-35, which had already been mass-produced and entered service with the U.S. Army Air Corps.

The XP-41 was the final aircraft produced in the P-35 program and rolled off the assembly line in 1938. The aircraft had a radial engine, similar to the P-35 design, but had numerous improvements that were considered to be unconventional in American fighter design at the time. These features included a streamlined canopy that enclosed the cockpit, retractable landing

The Seversky XP-41 ready for flight-research studies by the NACA at Langley in August 1939. The flight studies were performed to validate drag cleanup and engine-cooling tests carried out in the full-scale tunnel. Although the XP-41 never made it into production, data derived from the XP-41 program contributed to the development of its famous relative, the Republic P-47 Thunderbolt. (Photo courtesy of NASA)

The XP-41 photographed as it was rolled through the entrance door of the full-scale tunnel and parked in the return passage of the full-scale tunnel awaiting preparation for testing. Whenever possible, the NACA supported the military with combined wind-tunnel and flight evaluations of new aircraft. (Photo courtesy of NASA)

gear, and a turbosupercharged engine that provided considerable additional horsepower over the P-35 engine. The aircraft, which flew for the first time in March 1939, was capable of a top speed of 323 mph at 15,000 feet, and later benefited from drag-reduction modifications suggested by Langley.

The NACA drag-reduction studies on the XP-41 were part of a famous series of tests conducted in the Langley Full Scale Tunnel in what was referred to as "drag cleanup." This investigation method, which ultimately benefited every American fighter design during World War II, involved a process in which the full-scale aircraft was mounted in the tunnel, antennas removed, fairings and gun ports taped over, canopies sealed, and gaps and

cracks taped up. In the sealed and faired condition, the aircraft was then thoroughly examined for sources of drag. The sealing tape was removed over each individual part or section of the aircraft, one at a time, while researchers measured the drag produced by each step of the removal process. Data obtained from the tests were quickly provided to the military and aircraft manufacturers for immediate use, and subsequently in official NACA reports that also proposed ways of reducing the aircraft drag and enhancing performance. The full-scale tunnel tests usually were supplemented with flight studies of the aircraft to confirm the findings and suggestions for improvement.

The first drag clean-up test in the full-scale tunnel occurred in April 1938, when the Navy's Brewster XF2A-1 prototype underwent detailed analysis for drag

The Seversky XP-41 undergoing drag clean-up studies in the full-scale tunnel in 1939. The drag clean-up studies led to marked improvements in the aircraft's performance capabilities. Note how the aircraft's engine has been faired over and streamlined as a starting point in the drag-evaluation process. (Photo courtesy of NASA)

Close-up of the engine cowling, air scoops, and gun mounts on the nose of the Seversky XP-41. These areas usually produced the largest drag increments in clean-up studies and were given intense attention during entries in the full-scale tunnel. (Photo courtesy of NASA)

reduction. Recommendations for modifications were forwarded to the Navy with a prediction that the top speed of the modified prototype could be increased from 250 mph to 281 mph, an increase of more than 10 percent.

The Langley drag clean-up studies of the XP-41 design began in June 1939. During the tests, drag was measured for the engine and accessory cooling systems, the gun installation, radio antenna, engine cowling, canopy, and production abnormalities in the wings. The aircraft design's lift characteristics and stalling were also studied. The results of the tunnel tests revealed that an impressive level of drag was produced by a number of aircraft features. For example, the drag increments caused by the powerplant installation (cowling and cooling airflow, carburetor air scoop, accessory cooling, exhaust stacks, intercooler, and oil cooler) increased the drag of the sealed and faired condition by more than 45 percent. Additional drag caused by the gaps in cowling flaps, antenna, walkways, landing-gear doors, and gun-blast tubes was also very large. The combined drag of these additional items as well as the powerplant drag inducers increased drag by a very impressive 65 percent. An important lesson learned in this test was that although most airframe items produced individual drag increments of only a few percent, the sum of the increments added up to an impressive total increase in aerodynamic drag.

After analyzing the magnitude and causes of the incremental drag caused by aircraft components, the NACA researchers suggested improvements that significantly increased the XP-41's performance. For example, the powerplant drag could be reduced from 45 percent to 27 percent, and the roughness and leakage drag could be reduced from 20 percent to only 2.5 percent.

While the NACA Langley drag clean-up studies of the XP-41 led to marked improvements in the aircraft's performance, the design faced stiff competition from other competitors for the role of primary pursuit aircraft in U.S. Army Air Corps service. These competitors included the Lockheed XP-38 Lightning, Bell XP-39 Airacobra, and Curtiss XP-40 Tomahawk. The relatively cheaper XP-40 was chosen by the Army Air Corps to serve as the pursuit mainstay. The Army was, however, interested in another Seversky product, the YP-43 Lancer, and dropped interest in the XP-41.

Although the XP-41 never made it into production, it served as a research testbed that contributed a broad knowledge base for future fighter designs. It also served as an important example of the potential benefits of drag clean-up analysis for aircraft configurations. The results of the XP-41 studies in the full-scale tunnel have become textbook examples for drag-improvement studies. During the 1970s, NASA published a summary of drag clean-up results obtained by the NACA on 23 aircraft for use by the general aviation community, and the XP-41 study was a highlight of the summary report.

Curtiss XP-46

In September 1939, the U.S. Army Air Corps requested that Curtiss-Wright undertake the construction of a fighter aircraft designated the XP-46 that was more advanced and would supersede the Curtiss P-40. The XP-46 design was to incorporate design characteristics seen on pursuit aircraft in Europe considered to be "innovative" and highly desirable. Only two XP-46s were built. The aircraft possessed a retractable landing gear that folded up inward, had an Allison V-1710-39 engine, and was to be armed with two 0.50-caliber machine guns in the nose and as many as eight 0.30-caliber machine guns in the wings.

At Langley, wind-tunnel tests of the XP-46 were conducted in the full-scale tunnel in late 1939 at the request of the Materiel Division of the Army Air Corps. The goal was to study the best aircraft configuration for maximum high-speed performance with sufficient engine cooling. Stalling characteristics, as well as aileron and elevator-control effectiveness, were also investigated. The full-scale tunnel tests showed that the best aerodynamic performance was obtained with the addition of a large fuselage/wing fillet and a new wing leading edge. Researchers also found that a long-fuselage-nose configuration showed better maximum lift capability than the short-nose version. In addition, it was found that the radiator on the underside of the aircraft, with inlet scoop and outlet flaps open, degraded maximum lift.

The first XP-46 flew for the first time on February 15, 1941. The results of the flight tests revealed, however, that the aircraft could perform no better than the most advanced version of the P-40 at the time. Consequently the XP-46 program was cancelled. Many of the design characteristics of the XP-46, including the retractable landing gear, became standard characteristics of advanced fighters in the American arsenal in later years.

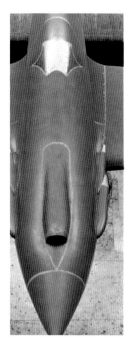

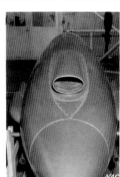

Curtiss XP-46 mounted in the full-scale tunnel for aerodynamic performance and engine-cooling studies in 1939. The studies showed that a long-nose fuselage configuration had higher maximum-lift capability than the short-nose version. Langley researchers also found that the radiator on the underside of the aircraft, with inlet scoop and outlet flaps open, degraded maximum lift. (Photo courtesy of NASA)

Close-ups of the Curtiss XP-46 with various cooling-duct intakes, as tested in the full-scale tunnel. (Photo courtesy of NASA)

Close-ups of an inlet-scoop modification showing fairing work to improve the XP-46's aerodynamic characteristics at high-lift conditions. (Photo courtesy of NASA)

CHAPTER 2

THE WAR YEARS

With America's entry into World War II in December 1941, the nation faced many challenges in forging a formidable aerial arsenal capable of achieving victory over the Axis air forces. Early Axis triumphs and global advances early in World War II were due largely to superior airpower. However, the tide slowly began to turn in favor of the Allies, primarily due to the industrial strength and ingenuity of aircraft manufacturers and designers in America. The United States was hard pressed to match Axis technological developments in Europe and Japan, but eventually caught up with these countries and along the way developed a superior fleet of advanced aircraft designs that helped win the war.

America's aeronautical research capabilities at the beginning of the war were becoming more advanced, and the priorities of supporting the war effort rapidly accelerated the technical and human resources at Langley. The total number of employees at Langley increased from 739 in 1940 to 3,288 in 1944. Critical tests and research in support of the military brought an overload of work to Langley's aeronautical research facilities, many of which operated on around-the-clock schedules, driven by the patriotism and innovation of their dedicated staffs. In recognition of the value of the NACA research on new designs, the military instituted an arrangement whereby the NACA was supplied with the third aircraft off the production line (later changed

Curtiss XP-42 used in flight research studies at NACA Langley in February 1942. For the test program depicted in the photograph, the aircraft was finished with a radial-engine, low-drag cowling and an engine-cooling fan that was inspired by European applications for fighter aircraft such as the German Focke-Wulf FW-190 design. (Photo courtesy of NASA)

to the fifth production airplane) for every new military aircraft. Wind-tunnel drag clean-up, stability and control, and spinning tests were high-priority activities, making up the bulk of studies at the full-scale tunnel, spin tunnel, and free-flight tunnel during the war. Radical new fighter and bomber designs were common sights at the facilities. Throughout the 1940s, however, many radical and innovative aircraft designs failed to make it into production. Designers were pushing the envelope in terms of higher top speeds, lower landing speeds, more effective controls, and overall efficiency and lethality.

Curtiss XP-42

The XP-42, originally the fourth Curtiss P-36A produced, was received by the U.S. Army in March 1939 as an attempt to incorporate a more streamlined engine cowling to provide reduced drag and improve performance for fighters powered by radial engines. Langley researchers also used this modified aircraft in a broad variety of NACA experiments, including an extensive evaluation of experimental engine cowling shapes to improve aerodynamic performance and

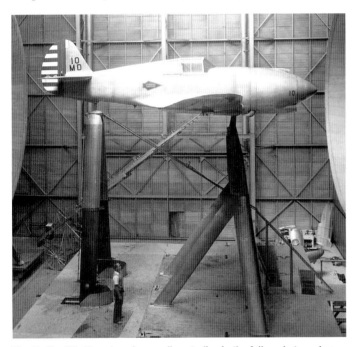

The Curtiss XP-42 undergoing cowling studies in the full-scale tunnel on September 14, 1940. The aircraft is in its original cowling configuration with the air scoops on the top and bottom of the cowling. In later tunnel and flight tests, the aircraft was outfitted with four different cowlings. (Photo courtesy of NASA)

cooling. The program, which included wind-tunnel and flight-test evaluations, was a large effort involving personnel from the NACA, Curtiss-Wright, Republic, Pratt & Whitney, as well as the program sponsor, the Army Air Forces' Air Materiel Command.

At Langley, the XP-42 was tested in the full-scale tunnel in September 1940 to investigate the cooling and aerodynamic performance of several engine/cowling concepts for aircraft having long-nose radial engines. The XP-42 used a Pratt & Whitney engine with an extended propeller shaft that placed the propeller about 20 inches further forward than the standard position. The forward extension of the propeller enabled the use of fuselage nose shapes of higher fineness ratio than those of the standard short-nose configuration. The original Curtiss-designed XP-42 used a pointed fuselage nose with sharp-edge air scoops at the top and bottom of the cowling for cooling. However, flight tests showed that the top speed of the airplane was no better than the short-nose P-36 airplane. NACA analysis and preliminary flight measurements indicated that the cowling scoops and cooling-flow characteristics caused excess drag. In particular, it was found that the engine-cooling air entered the lower air scoop at only about half the airplane's flight speed, and that the energy of this flow was rapidly dissipated by a sharp change in the airflow direction at the rear of the scoop to a large area of the engine. This large internal energy loss due to cooling flow led to a request for full-scale tunnel studies to identify the sources of internal and external drag and suggest improvements for the design.

The goals of the test program were to reduce external drag and increase the critical Mach number over the nose by reducing the negative pressure peak associated with high-speed flight, and to reduce cooling drag by increasing the cooling-air pressure recovery. Four modified cowling and cooling designs were evaluated, and it was demonstrated that a long-nose engine configuration such as this aircraft used could be modified with a very efficient annular inlet cowling. Langley had designed a promising new cowling shape referred to as the NACA "D" cowling that used an annular inlet and a diffuser section for engine-cooling air. The test program also determined that the critical speed for the onset of compressibility on the engine cowling could be increased to more than 500 mph at altitude with the modified cowl design. The investigation also revealed that significant drag losses occurred for the XP-42 due to the flow of cooling air out of conventional cowling outlets impinging on flap gear and exhaust collectors that disturbed the airflow. The full-scale tunnel tests were followed by an extensive series of flight tests at Langley from May 1941 to December 1942 to evaluate the effectiveness of several experimental cowling designs.

The XP-42 aircraft also played a key role at Langley in the development of technology for improving control of high-speed aircraft. Early experiences with compressibility effects encountered during high-speed dives of certain military aircraft had caused considerable

Close-up of the all-movable horizontal stabilizer flown on the XP-42 in a NACA research investigation on lowering control forces during high-speed dives. Note the trailing-edge tab used to assist in deflecting the tail. This concept significantly reduced control forces by placing the aerodynamic center of the movable surface in a more favorable location than for a conventional stabilizer/elevator combination. Tabs do not work well at supersonic flight conditions, and modern aircraft use hydraulic systems to deflect all-movable tails. (Photo courtesy of NASA)

Curtiss XP-42 in flight over Langley with all-movable horizontal stabilizers. With this modification the aircraft demonstrated excellent maneuverability and low stick forces. The all-movable horizontal-stabilizer concept is routinely used today for high-speed aircraft. (Photo courtesy of NASA)

alarm within the military services and industry design teams. The control forces required to recover from these high-speed dives using conventional aircraft elevator deflections were excessive and beyond the strength of many pilots. Langley researcher Robert R. Gilruth proposed a new type of longitudinal control in the form of an all-movable horizontal stabilizer controlled by an auxiliary tab. With this arrangement the control forces were significantly reduced during dives. In flight experiments conducted by Langley in 1941 and 1942, the XP-42 was fitted with an all-movable horizontal stabilizer and performed admirably, demonstrating excellent maneuverability.

The XP-42 proved to be slower than the Curtiss XP-40, which was the leading candidate for the mainstay Army fighter. As a result of its lack of comparable performance and the intense interest in the XP-40, the XP-42 never entered production. Nevertheless, the important technologies advanced by the research on this aircraft provided designers with options for radical concepts that significantly improved future aircraft. The cowling research provided fundamental design guidelines later used to a great extent on fighter aircraft, and the demonstration of the advantages of all-movable horizontal stabilizers as a design option for high-speed aircraft was a significant milestone in aviation history. All-moving horizontal tails are now a standard feature on all modern high-speed fighter designs.

General Motors "Flying Torpedo"

Before America's involvement in World War II, the General Motors Co. designed and built an unmanned flying bomb known as the GMA-1 Flying Torpedo for the Army Air Corps. The flying torpedo used a 200-hp piston engine and would take off from a four-wheeled dolly. The unmanned aircraft was capable of delivering a 500-pound bomb load with an operational range of 400 miles. It was automatically steered by means of elevator and rudder controls regulated by an altimeter and directional gyro. The drone's wingspan was 21 feet, and with the 500-pound payload it weighed 1,250 pounds. The Army placed an order for 10 experimental versions in 1941, and requested that the NACA undertake aerodynamic performance, stability, and control studies of the drone in the Langley Full Scale Tunnel in August 1941.

In the full-scale tunnel tests, Langley researchers investigated control effectiveness and determined the most efficient trim settings for the aircraft. Researchers also studied propeller/slipstream effects on the drone's stability and control with power on. In addition, projections were made regarding the drone's inherent stability and probable flight behavior. The results of the investigation indicated that the stability and control characteristics of the aircraft would be satisfactory and that a maximum speed of about 198 mph could be achieved with the piston engine. However, crashes associated with the launch and guidance mechanisms

Side view of the General Motors GMA-1 Flying Torpedo mounted in the full-scale tunnel for aerodynamic testing on August 20, 1941. The studies included assessments of the aerodynamic performance, stability, and control characteristics of the unmanned flying bomb. Trim settings were determined for maximum range. (Photo courtesy of NASA)

Front view of the GMA-1 Flying Torpedo during tunnel tests. During the wind-tunnel studies, Langley researchers determined that the drone possessed "good" stability and control qualities. However, the flying bomb's flight-test program was plagued by several crashes that resulted in the termination of the project in 1943. (Photo courtesy of NASA)

of the drone system plagued the test-flight program, and the Flying Torpedo program was subsequently terminated in 1943.

Vought V-173 "Flying Pancake"

As discussed in the previous chapter, the Guggenheim Safe Aircraft Competition of 1929 had triggered the goal of stimulating the growth of civil aviation by achieving designs for safer aircraft. This interest became even more widespread and carried over into the 1930s to professional aircraft designers, engineers, and even to inventors with no significant aeronautical credentials. As previously discussed, Fred Weick and a group at Langley had initiated an informal NACA competition to identify safer airplane designs. One of the participants in the activity was Charles H. Zimmerman, the brilliant Langley engineer who led the design of Langley's spin- and free-flight–tunnel facilities.

In the late 1920s, inventors had proposed several unorthodox disc-shaped civil aircraft configurations. In 1926, Dr. Cloyd L. Snyder, a podiatrist from South Bend, Indiana, was intrigued by the flying ability of a shoe heel lift insert that he had tossed into the air. Inspired by the ability of the odd shape to glide, Snyder thrust

Dr. Cloyd L. Snyder's Arup S-2 aircraft in flight with Glenn Doolittle at the controls. The configuration was powered by a 37 hp engine driving a tractor propeller. Weighing 780 pounds and with a wing area of 211 square feet the S-2 had a very low wing loading and could land at 23 mph. Note the wingtip aileron surfaces that provided roll control and the deflected elevator surfaces on the outer rear of the wing. A conventional rudder supplied yaw control. This photo, which shows the aircraft in early markings, was presented to the Langley staff during a demonstration visit in January 1934. (Photo courtesy of NASA)

himself and his resources into making the radical heel-shaped configuration into an airplane. Aided by an engineer and the fabrication of a small wind tunnel, Snyder built unpowered gliders and three piloted versions of aircraft designated the Arup (Air up) S-1, S-2, and S-4. With pilot Glenn Doolittle (James H. "Jimmy" Doolittle's cousin) at the controls, these "flying pancake" aircraft made hundreds of accident-free flights in the 1930s and impressed observers with low landing speeds. In one demonstration at the South Bend Airport, Doolittle flew the 780-pound Arup S-2 aircraft at speeds up to 97 mph on 37 hp and landed at 23 mph. In addition to flight demonstrations, Snyder used the aircraft for billboard-type advertising for customers that included Sears and Roebuck.

The most innovative feature of the Snyder aircraft was the use of a low-aspect-ratio wing. Aspect ratio is defined as the ratio of the wingspan dimension squared divided by the area of the wing. Conventional wisdom at the time was that the value of a wing's aspect ratio should be about six for optimum efficiency. The aspect ratio of the Snyder wing was much smaller (less than two) because of the relatively large wing area.

Meanwhile, in his pursuit of new technology for safer airplanes Zimmerman was interested in aircraft that could descend along a steep path at a low rate of speed and permit the pilot to land in almost parachute fashion without altering the flight path or speed. Such a capability would demand a relatively low value of lift-to-drag ratio and a high resultant force. Zimmerman was aware that low-aspect-ratio wings might satisfy both

requirements for the "parachute" effect he desired for landings. He was also aware of the Arup configurations and their highly successful flight demonstrations. He became interested in the aerodynamic principles behind low-aspect-ratio wing configurations and began a series of experiments and analyses at Langley to optimize the advantages of the design concept. In 1933 and 1935 he published reports on results of exploratory wind-tunnel tests he had conducted in the Langley 7-by-10-Foot Wind Tunnel on the aerodynamic characteristics of a series of low-aspect-ratio wings (with aspect ratios from 0.5 to 3). He also varied the wingtips with concave and convex shapes as viewed from the front. The results showed that a circular wing of aspect ratio 1.27 could produce almost twice the maximum lift of a high-aspect-ratio wing of equal area. This lift augmentation occurred at a very high angle of attack on the order of 45 degrees, and the shape of the wingtip had a large effect on drag for angles of attack near maximum lift. Zimmerman was delighted with his experimental results, and he continued with free-flying model tests.

In January 1934 Glenn Doolittle flew the Arup S-2 to Langley for a flight demonstration and discussions with the Langley staff.

Zimmerman continued to explore the options for optimizing the low-aspect-ratio wing. He hypothesized that by using large wingtip-mounted propellers he could immerse the wing in a relatively high-energy flow to maintain lift and, perhaps, by rotating the propellers "down at the wingtip," counteract the effects of the wingtip vortices and thereby reduce the induced drag of

Snyder's Arup S-2 (background) and S-4 (foreground) at the South Bend, Indiana, airport. Note the difference in control-surface arrangements between the aircraft. The S-4 used a conventional cruciform tail for pitch and yaw control with skewed ailerons on the wing for roll control, whereas the S-2 used wingtip ailerons, rudder, and wing-mounted elevators. Also note the updated finish and markings on the S-2. (Photo courtesy of Larry Loftin)

The Arup S-2 at Langley during a flight demonstration for the NACA staff on January 22, 1934. Side view shows the deflection of the trailing-edge elevators. Note that the center section of the wing beneath the vertical tail was not deflectable. The aircraft's low-speed performance impressed the Langley researchers, especially Charles Zimmerman. (Photo courtesy of NASA)

the low-aspect-ratio wing in conventional flight. As an entry in the NACA 1933 competition for innovative light aircraft, his flying-pancake design was considered the winner based on originality and innovation, but his NACA managers were not interested in the concept, declaring it to be "too advanced" and "too much of a novelty." NACA programmatic support for further studies was not forthcoming. Nonetheless, Zimmerman continued to design and test his radical design concept during off-work hours at his home with varying degrees of success.

Encouraged by the results of exploratory wind-tunnel and glider tests and confidence in his design, Zimmerman constructed a 20-inch-span rubber band-powered model of his design in 1936. This "concept demonstrator" finally influenced Zimmerman's bosses at Langley into permitting him to seek a corporate sponsor for a full-scale, manned version. The Navy, of course, was extremely interested in radical new concepts that might dramatically lower the takeoff and landing speeds for operations from aircraft carriers at sea. Zimmerman subsequently found a sponsor for the design in the Chance Vought Aircraft Division of the United Aircraft Corp., which had produced aircraft for the Navy and was in search of a new innovative naval aircraft contract. Zimmerman left Langley in 1937 to join Vought.

At Vought, Zimmerman built and flew a model known as the V-162, which embodied his twin-propeller design, with twin vertical tails and a hinged aft wing section to provide pitch control. In March 1939 Vought submitted drawings to the Navy to stimulate interest in the "Flying Pancake," designated the V-173. With interest in the design becoming more intense, in October the Navy requested that the NACA evaluate the flying characteristics of a small-scale model of the V-173 in the Langley Free-Flight Tunnel.

Initially, Vought supplied a 32-inch-span balsa model of the V-173 for Langley to test, but the model was much too fragile for the rigors of free-flight testing, and a new 24-inch free-flight model was fabricated. Tests of this 0.086-scale model for power-off conditions were initially conducted in late 1939, followed by power-on testing.

The original V-173 design had no outboard horizontal tails, and used elevator surfaces on the rear of the fuselage that could be deflected symmetrically for pitch control and differentially for roll control. Using the free-

flight technique and the unpowered model, the researchers found that the original V-173 design was unflyable. The model's handling qualities were terrible, and crashes were frequent, primarily due to lack of lateral stability. The vertical tail surfaces had to be increased in size, and the model was longitudinally unstable for low-angle-of-attack high-speed conditions.

Zimmerman suggested several design changes, and when the power-on tests were conducted, the model was evaluated with larger vertical tails and auxiliary horizontal tail surfaces on the outer rear of the fuselage that provided both pitch and roll control. Skewed aileron surfaces on the rear edges of the wing (similar to those used by the Arup S-4) were found to be ineffective for roll control at moderate and high angles of attack, and were dropped in favor of differently deflectable horizontal tails (referred to as "ailevators"). This configuration resembled the configuration that Zimmerman ultimately proposed in his patent application in late 1940. With this geometric layout, the model could be easily flown, and the Langley researchers concluded that the V-173 airplane would have satisfactory characteristics. Results of the test program were formally transmitted to the Navy in April 1940.

In May 1940, the Navy awarded Vought a contract for a single V-173 research aircraft. Built of wood and fabric, it was referred to as the "Zimmer Skimmer." The Navy quickly requested assistance from the NACA in the development of the configuration, and the Navy's Bureau of Aeronautics formally requested wind-tunnel

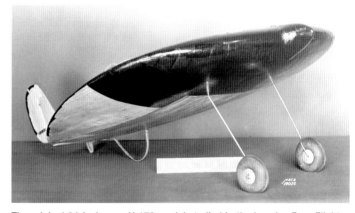

The original 24-inch span V-173 model studied in the Langley Free-Flight Tunnel. Note the lack of outboard horizontal stabilizers. Zimmerman had hoped that the shape of the wingtip would provide adequate lateral stability, but results showed otherwise. Many problems were identified, and fixes were declared mandatory before free-flight testing would be continued. (Photo courtesy of NASA)

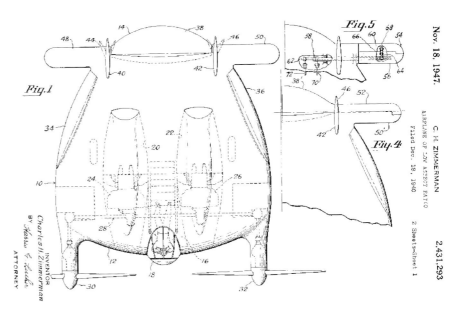

Copy of patent filed on December 18, 1940 by Zimmerman showing plan view of an "Airplane of Low Aspect Ratio." Horizontal tails were proposed for three possible options, including fixed surfaces to enhance stability, elevators for augmented pitch control, and differentially deflected surfaces for roll and pitch control ("ailevators"). Note the unswept horizontal tail surfaces and the skewed aileron surfaces on the rear of the fuselage. (Graphic courtesy of NASA)

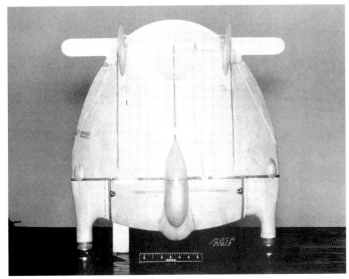

Refined Zimmerman configuration that emerged from Langley free-flight model tests. Note the fixed outboard tails and hinged roll-control surfaces on the trailing edge of the wing in front of the horizontal tails. Compare aileron surfaces with configuration of the Arup S-4 shown earlier. The roll-control aileron surfaces were evaluated in exploratory flight tests, but proved ineffective at moderate angles of attack and were dropped from the design. The horizontal tails were later swept back and made deflectable for the V-173 aircraft design. Results of the tests were reported to the Navy in March 1940. (Photo courtesy of NASA)

tests of the V-173 in the Langley Full-Scale Tunnel. The full-scale tunnel tests, which began in late 1941, were designed to determine the aerodynamic performance, drag, and airflow characteristics of the aircraft. One specific area of focus for the researchers was the aerodynamic efficiency of the configuration, especially the contributions of the aircraft's two huge propellers and the effects of the propeller slipstream on the characteristics of the overall configuration. The tunnel tests produced extensive aerodynamic data, including pressure distributions on the wing for various power-on conditions at combinations of angle of attack and angle of sideslip. The Langley engineers also investigated the effect of propeller rotation direction on lift and drag. For these tests the propellers (which rotated in opposite directions) rotated either "up-at-the-wingtips" or "down-at-the-wingtips."

The results of the test program provided some very surprising fundamental information on the aerodynamic performance of the radical configuration. Researchers found that, as had been expected, the inherently high induced-drag of a low-aspect-ratio wing could be partially compensated for by the favorable interaction of large-diameter propellers operating ahead of the wing. This effect was equivalent to an increase in the wingspan, since it resulted in increasing the mass of air to which downward momentum was imparted.

Prior to these tests, it had been conjectured by many that the secret of Zimmerman's design was to have the propellers rotate down at the wingtip, to oppose the rotational direction of the wingtip vortices and thereby minimize the tip losses and induced drag. However, the Langley tunnel test team unexpectedly found that the reduction in induced drag due to propeller operation was only affected to a small degree by the relative direction of the rotation of the propeller at the wingtip.

Vought V-173 "Flying Pancake" mounted in the full-scale tunnel for tests on November 28, 1941. Note the large two-bladed propellers used for the tests. The Short Takeoff and Landing (STOL) aircraft was tested for performance, stability and control, and to study aerodynamic interactions between the large propellers and the wing. Man in the photo is the lead NACA engineer for the test, John P. "Jack" Reeder, who later transferred from the full-scale tunnel to the flight research division and subsequently became one of most famous test pilots for the NACA and NASA. (Photo courtesy of NASA)

They also found that longitudinal stability was significantly decreased by propeller operation due to the direct lift on the forward-located propellers at high angles of attack, and the effect of propeller slipstream on downwash at the fuselage tail location. Another important result showed that changing the mode of

Top view of the Vought V-173 in the full-scale tunnel. During the power-on tests, Langley researchers measured and observed the aerodynamic characteristics of the aircraft's two huge propellers and the effects of their slipstreams on the overall configuration. (Photo courtesy of NASA)

Dramatic lower-front view of the V-173. Although the aircraft had been designed with "down at the tip" counter-rotating propellers to reduce the actions of wingtip vortices, the Langley full-scale tunnel tests showed little effect of the direction of propeller rotation on aerodynamic performance. (Photo courtesy of NASA)

propeller rotation from down-at-the-tip to up-at-the-tip resulted in a large increase in longitudinal stability with only a small decrease in performance. Finally, researchers used pressure distributions measured by instrumentation along the V-173's centerline to predict the critical speed at which undesirable compressibility effects would occur. While not practical for the underpowered V-173, the information was of interest for the high-speed follow-on fighter envisioned by Vought. The analysis indicated that the critical Mach number for the V-173 fuselage was 0.61, corresponding to an airspeed of 440 miles per hour at 15,000 feet.

When the rudimentary 2,250-pound (empty weight) V-173 first flew as a demonstrator of Zimmerman's concept on November 23, 1942, it confirmed the design's superb flight characteristics. With an aspect ratio of 1.27, it took off in still air at a speed of 29 mph, using only 200 feet of ground roll. The airplane was virtually impossible to stall or spin, and it favorably impressed several evaluation pilots, including Charles Lindbergh, who flew it in November 1943. The V-173 program ultimately completed 190 flights, with the aircraft last flying in March 1947. Since January 1942, Vought engineers were working hard on a true prototype full-scale fighter design with a weight of about 17,000 pounds. Performance objectives for the new fighter were a top speed of 460 mph, with a landing speed of 20 mph. In 1941, the Navy awarded a contract to Chance Vought for the development and production of the fighter, to be known as the XF5U-1. The Navy was particularly interested in the XF5U-1 as a candidate for interceptor duties aboard small carriers in the Pacific to help defend against the Japanese kamikaze menace.

The lightweight Vought V-173 concept demonstrator during a flight test. The V-173 demonstrated the excellent flight characteristics of Charles Zimmerman's design. The aircraft served as a developmental predecessor for the more advanced Chance Vought XF5U-1 fighter. (Photo courtesy of Larry Loftin)

Powered Chance Vought XF5U-1 model, minus cockpit and nose and engine intake installations, mounted in the full-scale tunnel for aerodynamic testing in 1945. Note the refined tail surfaces and configuration differences from the V-173. (Photo courtesy of NASA)

As part of requested support for the development program, a large 0.33-scale powered model of the XF5U-1 was first tested in the Langley Full Scale Tunnel in April 1945. The XF5U-1 configuration included four-blade articulated propellers rotating down at the wingtip and two all-moving outboard horizontal tails (called "ailevators") to obtain both longitudinal and lateral control. Circular inlets were incorporated into the leading edge of the wing for engine cooling, and a center cockpit was provided for the pilot. Hinged "stability flaps" located on the rear of the fuselage were used for longitudinal trim.

During flight tests of the V-173, pilots had reported a vibration problem that was caused by resonance between the cyclic propeller loads at high angles of attack and the nacelles. The problem was also encountered during taxi tests of the XF5U-1, and was expected to be very severe for in-flight conditions. Zimmerman then designed articulated blade hubs similar to the "teetering" hubs used by modern helicopters to mitigate the problem. The wind-tunnel model incorporated propeller blades that were free to flap fore or aft 10 degrees from the perpendicular to the propeller axis as they rotated. The blades were interconnected so that as one blade flapped forward, the opposite blade flapped aft. In addition, as a blade flapped forward, the propeller hub mechanism caused the propeller pitch to decrease, and as the propeller blade flapped aft, the pitch was increased. Zimmerman had conceived this load-relieving mechanism in an attempt to minimize large, unstable pitching moments expected by the operation of rigid propellers at high angles of attack. In addition, flapping blades might be used to optimize propeller stability, blade loads, and uniformity of thrust loading.

Front view of the XF5U-1 shows the engine intakes and characteristic stilt-like landing gear. Although it never flew, the XF5U-1 was one of the more radical aircraft designs of its day. (Photo courtesy of Smithsonian Institution via NASA)

The test program had originally included plans to determine aerodynamic characteristics of the large model with the articulated propellers removed and with the propellers operating. Tests of the model with the propellers removed included measurements to determine the longitudinal stability, the maximum-lift and stalling characteristics of various configurations, and the effectiveness of the control surfaces. Unfortunately, only a small part of the propellers' operating test program was completed when one of the model propellers failed during a test, resulting in complete destruction of the propeller. Therefore, the data obtained with propellers operating were very limited and insufficient to define the power-on stability and performance of the airplane. Nonetheless, interesting results were produced by this first test entry, including adverse results of mutual interference effects of the engine air-duct installations and a canopy on the wing on maximum lift. The effects of propeller operation on lift as measured in the brief power-on investigation were very large. About one-third to one-half of the total increase in lift due to propeller operation at high angles

Rear view of the Chance Vought XF5U-1 aircraft at the Chance Vought plant. (Photo courtesy of Smithsonian Institution via NASA)

of attack was due to the vertical component of the thrust of the propellers. At the time the tests were conducted, the projected gross weight of the XF5U-1 airplane was 16,750 pounds, and with the projected propulsive power a minimum forward-flight speed of about 90 mph was predicted.

Despite the STOL-potential of the XF5U-1 design, the end of World War II and the beginning of the jet age signaled the end of the XF5U-1 program. The Navy had become enchanted with jet interceptors, and interest in the XF5U-1 waned. As a result, the XF5U-1 program was cancelled in March 1947 with the aircraft never making a single flight. However, after repair, the XF5U-1 model had a second post-war entry in the Langley Full Scale Tunnel, in April 1948, to more fully test the effects of articulated propellers on the performance and stability characteristics of this STOL concept. Data collected would be used for future aircraft designs.

The results of the second test program confirmed Vought's expectations with regard to the effects of articulated- versus rigid-propeller operations. The destabilizing effects of propeller operation were much more pronounced for the rigid-propeller configuration; however, both propeller installations resulted in longitudinal stability under full-power conditions. The model could be trimmed using the ailevators alone to very high angles of attack of about 60 degrees for the rigid-propeller configuration, but no trim point was indicated for the articulated-propeller operation. Finally, the results showed that the stability flap could be used as a training device for normal-flight attitudes, but in transition flights to very low air speeds the trimming effectiveness of the flap was insufficient.

While the aircraft never made it into production, knowledge and data obtained from the XF5U-1 program led to improvements and refinements in future V/STOL concepts developed after World War II. In particular, testing techniques and knowledge gained from analysis of propeller/wing aerodynamic interactions was of great value when interest in powered-lift V/STOL configurations peaked in the late 1950s and early 1960s.

Both prototypes of the XF5U-1 were destroyed at the termination of the program. The V-173 was donated to the National Air and Space Museum of the Smithsonian Institution, and in 2003 it was transported to Dallas where it is now being renovated by retirees from Vought.

In 1948, Charles H. Zimmerman returned to Langley, where he continued his distinguished career with an immediate interest in V/STOL research. He stimulated many projects at Langley on radical concepts, from tilt-wing aircraft to flying platform devices. He was later selected to participate in the formulation of the NASA space program in 1958 and served as director of Aeronautical Research at NASA Headquarters in 1962-1963.

Republic XP-69

The Republic Aviation Company developed a radical fighter design designated the XP-69 for the Army Air Corps in the early 1940s. The configuration featured the unusual Wright R-2160 liquid-cooled radial engine, positioned to the rear of the cockpit (similar to the P-39 Airacobra) and driving contra-rotating three-blade propellers. The airplane, designed for high-altitude missions, also had a pressurized cockpit. Its weaponry consisted of two 37mm cannon and four 0.50-cal. machine guns.

The Army Air Corps Materiel Command requested that Langley conduct wind-tunnel tests of a powered 0.75-scale model of the XP-69 in the Langley Full-Scale Tunnel to study the longitudinal and lateral-directional stability and control characteristics of the aircraft. At the time, little data was available concerning the stability

During the XF5U-1 tests, an accident occurred when the starboard prop separated from the model and was destroyed. The wind-tunnel tests of the XF5U-1 confirmed the aerodynamic soundness of the advanced "Flying Pancake" design. (Photo courtesy of NASA)

and control behavior of aircraft with dual-rotating propellers, and the results of these tests were greatly anticipated by Army, NACA, and aviation officials. The large XP-69 model used sheet-aluminum metal outer skins that were filled and sanded to a smooth finish. Two 10-foot-diameter propellers were powered by two 25-horsepower electric motors in the fuselage. The dual-rotating propellers for the XP-69 had been previously tested in the Langley Propeller Research Tunnel. Slotted Fowler-type flaps were used as high-lift devices. During the tunnel tests, forces and moments on the model were measured at different angles of attack and sideslip. Tests were performed with the propellers off and with propellers running, and with the flaps in both the retracted and deflected positions. Propeller slipstream effects were also studied.

The results of the full-scale tunnel tests in August 1942 revealed that the design was longitudinally stable

The 0.75-scale model of the Republic XP-69 mounted in the full-scale tunnel for testing to determine longitudinal and lateral stability and control characteristics. Note the contra-rotating propellers, Fowler-type flaps, and underslung intake. (Photo courtesy of NASA)

through the full range of lift, but a significant reduction in stability occurred for power-on conditions because of the direct contribution of the propeller's normal forces forward of the center of gravity. The elevator control was sufficient for all tests conditions. The potential for dynamic lateral instability ("Dutch Roll") was identified because of excessive dihedral effect and low directional stability for flaps-down landing conditions at idling power. Roll control was low compared to contemporary fighter aircraft, and lateral stick forces at high speeds were predicted to be unsatisfactory. The tests also showed that there was virtually no asymmetry in yawing or rolling moments for cruise or flight at low angles of attack, and that no aileron or rudder input would be required to maintain an unyawed attitude during normal operations. This result was attributed to the relatively symmetrical slipstream flow (no rotational effects) behind the dual-rotating propellers.

Despite the relatively promising aerodynamic test results, teething problems were encountered with the XP-69's engine. As a result, the Army Air Corps lost interest in the XP-69 program and decided to encourage and pursue the development of the Republic XP-72 instead. The XP-69 program was terminated in May 1943 with only one aircraft ever being built.

Curtiss XP-55 Ascender

In late 1939, the Army Air Corps requested proposals for a new fighter-aircraft design from manufacturers, emphasizing the Army's interest in truly

Overhead view of the XP-69 model mounted in the tunnel shows its graceful lines. The Fowler flaps are deployed for high-lift tests. Langley found the configuration had low levels of longitudinal and directional stability for power-on conditions. (Photo courtesy of NASA)

radical configurations. The Curtiss-Wright Corp. responded with a unique, free-floating canard, pusher-propeller design with a swept-back wing. The aircraft, known as the XP-55 Ascender, also featured a tricycle landing-gear arrangement, which had been demonstrated by the Weick W-1A as discussed earlier in this book.

The XP-55 program began in June 1940 when Curtiss-Wright received the Army contract that called for the construction of a powered wind-tunnel model for assessments of the configuration. The contract deliverables were a wind-tunnel model, some preliminary wind-tunnel data, and an option for an experimental aircraft. After building and testing a large model of its design, the Curtiss-Wright team received a lukewarm reception from the Army based on the test results; but the company pressed on under its own funding with the design and construction of a flight-worthy, full-scale demonstrator aircraft known as the CW-24B. The fabric-covered, wooden-winged CW-24B was intended to be a lightweight, low-powered flying test bed for the full-scale fighter design.

Wind-tunnel development testing of the CW-24B configuration in 1941 included tests of a 0.25-scale model in a tunnel at the Massachusetts Institute of Technology and the NACA Langley 19-foot Pressure Tunnel. Both test programs identified an early concern regarding flow separation on the outer swept-wing panels at high angles of attack. The flow separation resulted in longitudinal instability ("pitch up") near stall. In mid-1941 the Materiel Command of the Army Air Forces requested that the NACA test a free-flying

Unpowered free-flight model of the early CW-24B configuration. Note the small inboard vertical tails. NACA free-flight tunnel results indicated that the full-scale airplane would be directionally unstable, have excessive dihedral, and insufficient aileron effectiveness. In addition, when the canard was fixed, the model became longitudinally unstable. Langley recommended moving the vertical tails to an outboard wing location. (Photo courtesy of NASA)

model of the CW-24B in the Langley Free-Flight Tunnel to evaluate its general flying behavior. By late November, a 0.10-scale flying model had been constructed and prepared for tests. In the first test entry the model was unpowered, followed by a second entry for powered tests in 1942.

The results obtained during flying model tests in the free-flight tunnel highlighted several problem areas. The tailless CW-24 incorporated vertical tail fins for directional stability, but the test results showed that the configuration was directionally unstable. The design was also found to have excessive geometric dihedral, and the ailerons were not sufficiently powerful. The longitudinal instability noted in the results obtained from the other wind tunnels was also present for this model. Finally, a special concern was cited over the longitudinal control system used for the free-floating canard. When the canard was allowed to float freely and align itself with the local flow at the nose, the longitudinal stability below stall was satisfactory. However, if the pilot flying the model held the elevator in a fixed position for even a few seconds, the model was longitudinally unstable to an alarming degree. Langley researchers recommended installing vertical fins at the wingtips, reducing the geometric wing dihedral angle, increasing the aileron area by 25 percent, and reviewing the longitudinal control-system design, which might cause stability problems for fixed-canard conditions in critical flight regimes. By the time the model entered the free-flight tunnel for the second time, it had been modified according to the NACA recommendations, and it also included changes that had been made to the full-scale aircraft, including engine cowl fins and revised wingtips. Results of the second entry indicated that the directional-stability problem had been solved with the vertical fins at the wingtips, but the difficulties previously encountered for excessive dihedral, lack of aileron control, and stick-fixed longitudinal instability remained.

Initial flight tests of the CW-24B were conducted at Muroc Bombing Range (now Edwards Air Force Base), California, in December 1941. The early wind-tunnel tests had indicated that, following a longitudinal instability at stall, the CW-24B configuration would exhibit a "deep stall" condition in which the airplane would trim at very high positive or negative angles of attack and descend in near-vertical flight in an uncontrollable condition with an almost horizontal attitude. However, the underpowered CW-24B demonstrator was not capable of gaining sufficient altitude to permit spin recovery, and intentional spins were not attempted in the uneventful preliminary flight evaluation, which included over 190 flights. In July 1942 the Army Air Corps awarded a contract for three operational XP-55 aircraft.

At the end of the CW-24B flight-test program in May 1942, the Army Materiel Command had acquired the aircraft from Curtiss and requested that Langley undertake a full-scale tunnel investigation of its aerodynamic characteristics. Quickly approved by the NACA, the tunnel

Side view of the CW-24B (concept demonstrator for the Curtiss XP-55 Ascender) mounted in the full-scale tunnel on October 27, 1942 for longitudinal stability and control testing. The CW-24B was designed as a lightweight, low-powered test bed with fabric covering and wooden wings. Note the free-floating canard and the outboard vertical tail location. (Photo courtesy of NASA)

tests began in October 1942. For these tests, the CW-24B was modified to more closely represent the evolving XP-55 configuration. The canard was modified to that of the XP-55, the previously fixed landing gear of the CW-24B was removed, and the wing was resurfaced to a smooth finish. Lift, drag, pitching-moment, hinge-moment, and elevator-pressure measurements were made for variations in angle of attack, wing-flap deflection, elevator deflection, and elevator-tab settings. All tests were made for propeller-removed conditions. Flow separation and wing-stalling behavior were also monitored by analyzing wool tufts on the wings of the aircraft.

A key feature of the CW-24B and XP-55 designs was the longitudinal control concept, provided by the all-movable, free-floating canard surface. The canard was directly connected to the pilot's control stick, and it also featured tabs that were controlled by a separate trim control in the cockpit.

Results of the full-scale tests correlated well with the earlier subscale model tests. Langley researchers found that the CW-24B was longitudinally unstable when the stick was fixed with the propeller removed. With the stick free (canard free-floating) and the landing flaps in a retracted position, longitudinal stability could be maintained at angles of attack below about 12 degrees, but the aircraft exhibited pitch-up at higher angles of attack. Researchers also found that a flap deflection of 45 degrees reduced the level of instability when the stick was fixed, and enhanced stability when the stick was free. Visualization results showed that flow on the rear

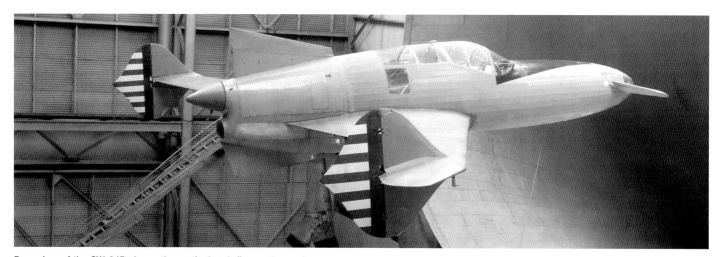

Rear view of the CW-24B shows the vertical stub fins on the engine compartment and engine exit. Note the outboard vertical tails, swept wingtips, and deflection of the free-floating canard. The full-scale tunnel tests showed that the design was longitudinally unstable for some conditions and that deflecting the wing trailing-edge flaps could reduce the instability. (Photo courtesy of NASA)

of the swept wing at angles of attack approaching wing stall (17 degrees) was in a spanwise direction and parallel to the trailing edge, promoting early flow separation and loss of lift behind the center of gravity, thereby resulting in the pitch-up tendency. This result was a precursor to the generic problem of longitudinal instability of high-aspect-ratio swept wings at moderate angles of attack.

The first XP-55 prototype flew for its maiden flight on July 13, 1943, experiencing an excessive takeoff run. Consequently, the nose elevator was increased in area by 15 percent, and an aileron up-trim connection to the flaps was implemented in an effort to bring the nose up earlier in the takeoff run. A major setback to the XP-55 program occurred on November 15, 1943, when test pilot Harvey Gray experienced an uncontrolled, inverted "deep stall" while conducting stall tests. Gray had successfully conducted two stall-recovery tests with flaps up, but when he attempted to evaluate the effects of wing spoilers on stall characteristics at idling power with the wing flaps and landing gear down, the aircraft pitched rapidly in a nose-down direction, past the vertical and onto its back in an inverted deep stall similar to that predicted by earlier tunnel tests. The aircraft locked into a flat, inverted stabilized condition as altitude was lost for more than 16,000 feet. Gray finally managed to bail out of the airplane safely, but the aircraft impacted the ground in an almost horizontal attitude and was destroyed.

In the wake of the accident, the Army Air Technical Service Command requested that Langley conduct wind-tunnel longitudinal trim tests of the XP-55 configuration at extreme angles of attack in the Langley 15-Foot Spin Tunnel to develop fixes or modifications to eliminate the unacceptable deep-stall behavior. Initially, Langley researchers tested a 0.059-scale model of the XP-55 on a special test rig that permitted a single-degree-of-freedom in pitch to establish correlation with the flight results and develop modifications to eliminate the trim conditions at extreme angles of attack. When it was placed in the tunnel's vertical airstream on the rig, the model would seek a trim angle of attack, and researchers would evaluate the stability of the trim condition by disturbing the model by pulling and releasing strings attached to the nose of the model. The model was provided with a number of candidate configuration modifications including alternate wingtips, small and large canards, wing leading-edge spoilers, wing fences

(small vertical fins on wing upper surface), and horizontal aft-cowl fins.

In its original configuration, the free-to-pitch XP-55 model would seek trim at very large positive (58-degree) and negative (70-degree) angles of attack, in qualitative agreement with the flight results and previous wind-tunnel results obtained with models of the CW-24B. Most of the candidate modifications, such as leading-edge root spoilers, wing fences, and elevator size, had no effect on the model's tendency to trim at large angles of attack. Large horizontal fins (auxiliary aft-horizontal tails) on the engine cowl prevented trim at large angles of attack, but were impractical from an operational perspective. The most effective cure for the deep-stall problem was found to be a larger canard with extended deflections (± 60 degrees), larger wingtips, and outboard ailerons called wingtip trimmers. The XP-55 model was then modified to include the recommended changes and tested for spin and recovery characteristics in the newly operational Langley 20-Foot Vertical Spin Tunnel. Langley researchers found that spins of the model could be stopped and recovery to conventional flight accomplished by rudder reversal and forward stick input. These results were later correlated with test results of a larger 0.20-scale model in Wright Field's Vertical Tunnel, and the third XP-55 aircraft subsequently was modified to incorporate changes to cure the deep-stall problem. This particular airplane was lost in a tragic fatal accident during an air show flyby at Wright Field in 1945. The design modifications were finally applied to the second aircraft, which was test flown by the Army Air Force

The restored second Curtiss XP-55 Ascender (serial number 42-78846) on exhibit at the Air Zoo, Kalamazoo, Michigan. (Photo courtesy of the Air Zoo, Kalamazoo, Michigan)

through 1944. Unfortunately, even with the aircraft modifications, stall characteristics of the XP-55 remained unsatisfactory. The stall occurred abruptly without warning, and rapid post-stall motions in pitch and roll were encountered. The airplane also had other disappointing characteristics, including deficient engine cooling.

The XP-55 exhibited disappointing performance compared to other front-line conventional fighters of the day. This factor, coupled with the emergence of the jet fighter, signaled the end of the XP-55 program. The aircraft has been remembered through history, however, for its radical design features, including its tricycle landing gear, swept wings and canard, and the use of a jettisonable propeller in the event the pilot had to bail out of the aircraft. The second XP-55 has been painstakingly renovated and is now on display at the Air Zoo Museum in Kalamazoo, Michigan, on loan from the National Air and Space Museum.

Northrop XP-56

Another radical aircraft configuration proposed for the Army's 1939 fighter competition was the Northrop XP-56 Black Bullet – a tailless pusher fighter with drooped outer wings and tricycle landing gear. It incorporated many of the same concepts used by the Northrop N1M flying-wing including turned-down wingtips for increased directional stability, wingtip rudders for directional control, and combined aileron and elevator control surfaces. Unlike the N1M, the XP-56 had a fuselage and vertical tail. Originally known as the N2B, the XP-56 featured three-blade contra-rotating

Free-flight model of the Northrop XP-56 Black Bullet. Results of Langley tests indicated that the design would have problems developing sufficient lift for acceptable takeoff characteristics. The design also exhibited insufficient directional stability. (Photo courtesy of NASA)

pusher propellers. The first prototype was ordered in September 1940, and a second prototype was requested in February 1942.

In early 1942, the Materiel Command of the Army Air Forces requested that Langley perform free-flight tunnel tests to investigate the stability and control characteristics of the configuration. A 0.083-scale model of the XP-56 was used for the tests, which included evaluations of several vertical-tail configurations and center-of-gravity locations. The results of the free-flight tests indicated that the configuration would exhibit excessive longitudinal stability for the forward center-of-gravity locations planned for the aircraft, and that without high-lift devices the aircraft would have a lower maximum lift capability than a competitive conventional airplane. The directional stability of the XP-56 with the original vertical tail configuration was found to be inadequate, and options for increasing the tail size were provided to Northrop. In addition, the split rudders were found to provide adequate yaw control. In general, the results showed that the XP-56 had good possibilities of being developed into a stable and controllable tailless airplane. However the directional-stability problem and high-lift concern remained following the first Langley test program.

Additional free-flight testing was conducted in mid-1942 to determine methods to enhance the aileron and elevator effectiveness of the XP-56 and to develop a high-lift device for the airplane. During the tests, the model was modified to maximize lift and enhance lateral and longitudinal control with the addition of leading-edge slots, various wing trailing-edge split flaps, and new elevons. The poor maximum lift and elevon effectiveness for the original design were dramatically improved by the modifications. The addition of leading-edge slots alleviated wing stall and enhanced aileron effectiveness and lateral stability through the low-speed regime. To improve the aerodynamic performance of the XP-56 design, Langley researchers ultimately recommended that new elevons be incorporated with greater surface area, that leading-edge slots be added to the wings, and that split flaps be used in the final version.

Flight testing of the Black Bullet was initiated at Muroc, California, in September 1943, at low altitudes. Tests of the first prototype came to an end, however, when the plane was lost due to a blown tire during a high-speed taxi. Northrop heeded the advice provided earlier by NACA Langley and incorporated split flaps in the second XP-56 prototype, which resulted in enhanced lift and directional control. In addition, an attempt was made to correct the longitudinal-stability issue by altering the center of gravity in the design.

When the second Black Bullet took flight in March 1944, it had been modified with an increased vertical tail size, air-bellow-operated rudder controls, and a modified center-of-gravity location. The aircraft's performance was still disappointing, and it was sent to the NACA Ames Aeronautical Laboratory for tests in the huge 40- x 80-foot wind tunnel in late 1944. Consequently, the XP-56 program languished before finally being terminated one year later.

Bell XP-77

As war clouds gathered in late 1941, the Army Air Corps became concerned over a possible shortage of aluminum and other critical aircraft fabrication materials. In October 1941, Bell Aircraft Corp. responded to a request by the Army for the design of an unconventional "light" fighter to be constructed from "non-strategic" materials. The resulting Bell design was a single-engine, low-wing configuration constructed almost entirely of wood and magnesium alloy that featured tricycle landing gear and a bubble canopy for enhanced pilot vision. Designated the XP-77, the aircraft was to be armed with a 20mm cannon and two 0.50-cal. machine guns. The little fighter was also to be capable of carrying a 300-pound bomb.

At Langley, wind-tunnel tests to measure and analyze the stability, cooling, and air loads of the XP-77

were carried out in the full-scale tunnel during two separate tests in June and October of 1943, at the request of the Materiel Command of the Army Air Forces. By that time, Bell had experienced considerable delays in the XP-77 program, and the Army ordered only two prototypes. The two airplanes were to have the same air-cooled engine, but different operational missions. The first version of the airplane was a low-altitude fighter with a design altitude of 12,000 feet, while the second version was a high-altitude fighter intended to operate at 27,000 feet. The full-scale mockup tested in the tunnel had a propeller thrust-line location similar to the low-altitude version of the airplane, but it used the 10.5-foot-diameter propeller intended for the high-altitude version. The landing gear was removed from the mockup for all tunnel tests.

Rear view of the XP-77 in the propeller-research tunnel. (Photo courtesy of NASA)

View of the Bell XP-77 mounted for tests in the Langley Propeller Research Tunnel in 1943. The XP-77 was fitted with a special NACA cowling whose shape was derived from propeller-research tunnel studies aimed at investigating the cowling and cooling limits of the Ranger SGV-770 engine. (Photo courtesy of NASA)

The tests' objectives were to determine longitudinal and lateral-directional stability, overall aircraft drag, the internal and external airflow qualities of a new NACA-designed cowling, and the air loads and critical speeds on both the cowling and the canopy. The NACA cowling was the product of extensive research that had been conducted with the same airplane in the Langley Propeller Research Tunnel earlier in the year to investigate the cowling and cooling limits of the Ranger SGV-770 engine.

The researchers conducting the full-scale tunnel tests found that, with power on, the cowling with a modified cooling-air exit area provided "excessive" engine cooling for cruise, but the cooling was found to be inadequate for a climbing attitude. As a result,

The Bell XP-77 undergoing aerodynamic testing in the full-scale tunnel in June 1943. During the tests, drag was studied as well as the internal and external airflow qualities of a NACA-designed cowling installed on the aircraft. (Photo courtesy of NASA)

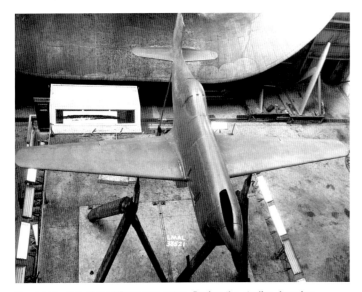

Overhead view of the XP-77 test setup. During the studies, Langley researchers recommended that the engine cooling-air exit area be enlarged or that cowl exit flaps be incorporated in the cowling design for adequate engine cooling. (Photo courtesy of NASA)

Langley recommended that the cooling-air exit area either be enlarged or that exit flaps be used in the cowling design to facilitate adequate engine cooling. Critical speeds for onset of compressibility for the cowling and the canopy were determined, with a speed of 415 mph being calculated for flow over the cowling at 12,000 feet and a speed of 496 mph for the windshield and canopy. The maximum lift coefficient of the aircraft was determined, and no discrepancies were found when varying the angle of attack or running the propeller. However, the engine failed while undergoing further tests, forcing the termination of the studies with no additional recommendations being made to enhance the aerodynamic performance of the aircraft. The results of the tunnel tests to determine stability and control characteristics did not disclose any serious problems.

Continued setbacks in the development program, including problems managing an appropriate weight of the design, forced planners to wait until 1944 to test fly the aircraft. Two prototypes were built, and on April 1, 1944, the first XP-77 prototype took to the skies for the first time over Wright Field. The initial flights exposed vibration difficulties associated with the engine/airframe layout. In addition, test pilots complained that they could not see very well from the cockpit due to the long nose of the airplane and the placement of the cockpit on the fuselage. The aircraft was reportedly hard to fly and was severely underpowered (the XP-77 had an unsupercharged XV-770-7 engine). Testing of the second XP-77 prototype at Eglin Field resulted in the destruction of the aircraft when the pilot unsuccessfully attempted an Immelmann maneuver that evolved into an uncontrollable inverted spin, forcing him to bail out. In December 1944, the XP-77 program was officially cancelled, with no production models being built.

Northrop MX-334

While Germany was pursuing the development of its Messerschmitt Me 163 Komet rocket-powered interceptor, America was quietly developing the technology as well. Northrop began the American effort in September 1942 when it investigated the potential of such an aircraft. These conceptual studies resulted in the company winning a contract for the development of three gliders. Two of the unpowered gliders became known as the MX-334, while the third, designated the MX-324, was slated for rocket-powered tests.

These aircraft were designed as experimental aircraft to thoroughly investigate stability and control characteristics of the radical design. The gliders, essentially flying wings, were primarily of wooden construction, with the exception of a metal center section. The pilot lay prone during flight so as to eliminate aerodynamic drag created by a conventional canopy. For improved stability, Northrop later decided to add a vertical fin to the design.

At the request of the Army Air Forces Materiel Command, Langley conducted full-scale tunnel tests of the MX-334 in 1943 to study the longitudinal and lateral stability and control characteristics of the all-wing glider. The NACA researchers were particularly interested in the aerodynamic performance of the design, which had neither a conventional fuselage nor vertical tails. These tests also included investigations aimed at identifying an appropriate wingtip, leading-edge, and slat configuration capable of enhancing the glider's static longitudinal stability and maximum lift capability. In addition, the program included a drag analysis and an evaluation of directional stability with vertical fins incorporated in the design. The effectiveness of the unconventional air-operated directional control system (similar to that used by the XP-56) was also evaluated. In this system, the inboard sections of the trailing-edge surfaces were operated by air bellows and provided both dive braking and directional control.

During the tunnel tests Langley researchers discovered that the aircraft, with slats removed, experienced tip stall of the swept wing at high angles of attack, resulting in a serious pitch-up problem not unlike that exhibited by the XP-55. They also found that the original wing slats proposed by Northrop were not as effective as desired in eliminating the deficiency. The addition of large-span slats maintained attached airflow at the tips until after the center wing section had stalled, thereby eliminating the pitch-up problem while enhancing lift. Also, the incorporation of vertical fins at the wing center section proved to enhance lateral-directional characteristics at high angles of sideslip. However, researchers also found that the aircraft's unorthodox directional control methods were insufficient. Rather than using the designed air-operated bellows to operate inboard elevons for yaw control, the NACA test team recommended that the duct system be modified if the existing directional control system was employed. Further recommendations included detailed modifications to the inlets, ducting, and butterfly control valve.

Flight testing of the MX-334 began on October 2, 1943. These tests, in which the glider was towed behind

Overhead view of the MX-334 clearly shows the leading edge slats. During the full-scale tunnel tests, vertical fins were added to the wing center section, which proved to enhance lateral-directional characteristics at high angles of sideslip. (Photo courtesy of NASA)

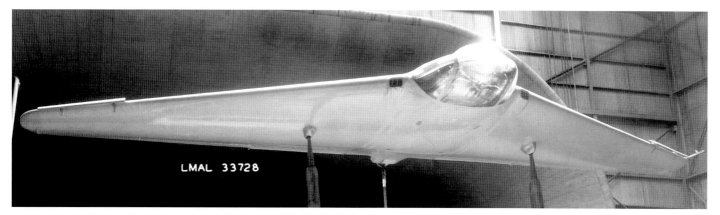

Northrop MX-334 glider mounted for tests in the full-scale tunnel on July 13, 1943. The tests were conducted to study the configuration's longitudinal and lateral stability and control characteristics. The tests also icluded evaluations of the wingtip leading-edge slat configuration shown in the photograph. (Photo courtesy of NASA)

a P-38 tow plane, proved to be hazardous. On one of the flights, the MX-334 was severely affected by the prop wash of the tow aircraft and became uncontrollable. The little glider entered a spin and eventually regained stability, although the aircraft was inverted and uncontrollable in an apparent inverted deep stall condition. While the pilot was able to parachute to safety, the out-of-control glider descended to earth in a series of circles and was destroyed on impact.

Flight testing of the rocket-powered MX-324 proved to be more successful. On July 4, 1944 the rocket-powered glider took to the air. Following release from the P-38 tow aircraft, the pilot performed a near flawless flight in which the glider remained in the air for a little over four minutes. No further tests of the aircraft were conducted and no production variants were ever produced; however, the aircraft contributed much toward the advancement of American aviation. The little flying wing had laid the foundation for data and knowledge that would be drawn upon for the development of the XB-35 and YB-49 designs, forerunners of today's highly successful B-2 Spirit stealth bomber.

Boeing XB-39

The XB-39 "Spirit of Lincoln" was an experimental variant of Boeing's highly successful B-29 Superfortress

Boeing XB-39 engine/wing section installation, ready for cooling tests in the full-scale tunnel on January 11, 1944. The tests were conducted to verify that design modifications to the engine powerplant had corrected a cooling problem detected in ground tests. The full-scale tunnel has been used for numerous power-on engine cooling and drag-reduction studies. (Photo courtesy of NASA)

Front view of the instrumented XB-39 engine powerplant used in static tests in 1944. Conducted as a back up for the XB-29 program, the XB-39 was cancelled when the XB-29 proceeded. (Photo courtesy of NASA)

Front view of the XB-39 test setup in the full-scale tunnel. (Photo courtesy of NASA)

A fuel-farm installation outside the full-scale tunnel provided capabilities for outdoor engine static tests. In 1944 static tests of the instrumented XB-39 engine were conducted at the site. (Photo courtesy of NASA)

heavy bomber. The XB-39 was actually the first YB-29 produced, but had inline liquid-cooled engines rather than the conventional air-cooled radial engines. It was the intent of Boeing and Army Air Force officials to use the XB-39 as a backup in case the conventional B-29 design with Wright R-3350 radial engines encountered developmental problems. Only one XB-39 was produced, and it was extensively tested in 1944.

Ground testing of the XB-39's Allison V-3420-11 powerplant (engine-nacelle configuration) revealed engine-cooling problems. Consequently, the engine was modified. The Army Air Forces Air Technical Service Command requested that Langley conduct wind-tunnel studies of the aerodynamic and cooling characteristics of the modified XB-39 engine installation in the Langley Full Scale Tunnel. These wind-tunnel tests were conducted from October 1943 to January 1944 to verify that the design modifications to the engine powerplant had corrected the cooling problem. The full-scale tunnel investigation included tests of the exhaust-shroud system and four modifications of the shroud system devised while running the tests. During the tests, Langley researchers studied the coolant, oil, and charge-air cooling systems with the propeller removed and in power-on tests. Researchers also analyzed the drag of the engine nacelle and cooling flap. Static tests of the instrumented powerplant were also conducted at an outdoor ground test site near the full-scale tunnel in 1944.

Results of the NACA tunnel tests showed that the original shroud system would be inadequate for cooling purposes during a simulated climb at 35,000 feet. With the additions of modifications to the shroud system recommended by the NACA, it was estimated that adequate cooling could be achieved at all altitudes and flight conditions. In addition, intercooling inadequacies were also identified. To correct these inadequacies, researchers recommended that the intercooling air outlet be reshaped and the outlet area be enlarged. They also recommended that coolant radiators with larger frontal and exit areas be used to facilitate better engine cooling. In addition, it was found that when the intercooler flap was fully opened, a reduction in lift was incurred.

Development of the B-29 proceeded in parallel with the XB-39. However, the B-29 program proceeded without any "show-stoppers," and the XB-39 program fell into obscurity with no production models ever being produced.

Free-flight model of the Kaiser Cargo Wing configuration studied in the Langley Free-Flight Tunnel. The original configuration had vertical tails mounted on the wingtips, but the Langley researchers predicted that the design's directional stability was unsatisfactory. Note the four vertical tails that the NACA recommended for satisfactory directional stability and the wire guards on the leading edge of the wing for protection during crashes in tunnel flight testing. Flying characteristics were satisfactory for the modified configuration. (Photo courtesy of NASA)

Kaiser Cargo Wing

During World War II, America's merchant shipping lifeline to Britain was being ravaged by the German U-Boat menace. To avoid this threat, industrialist Henry Kaiser proposed building large cargo-carrying flying wings capable of trans-Atlantic flight. The Kaiser Cargo Wing design was to feature four piston engines located in the front center section of the wing driving 15-foot-diameter four-blade propellers, with four vertical fins located behind the four engines. The cockpit was to be located atop the center section of the wing. The aircraft had a wingspan of 290 feet, a wing area of 7,920 sq feet, and a gross weight of about 175,000 pounds.

A 0.017-scale model of the huge aircraft was tested in the Langley Free-Flight Tunnel in 1943 at the request of the Navy. Initially, the design had only wingtip mounted vertical tails, but Langley researchers quickly determined that the directional stability of the configuration would be unsatisfactory as based on the free-flight model tests. Inboard vertical tails were recommended, and Kaiser designed four tail surfaces that were accepted for the modified Kaiser Cargo Wing configuration. With the modified tail configuration, the free-flight model displayed satisfactory flying characteristics.

As a follow-on investigation of interest to the Navy and the NACA, Langley researchers conducted tests of the same model in the NACA spin tunnel to determine its susceptibility to tumbling. In these tests, the model was released from a nose-up attitude in the tunnel's vertically rising airstream to simulate an attitude that might be reached during a whip stall. In these tests, the model would not tumble, but instead would quickly damp its motion to a vertical dive, indicating that the full-scale airplane would not be susceptible to the tumbling phenomenon that had been exhibited by several other flying-wing configurations.

A large wooden 0.143-scale powered model of the Kaiser Cargo Wing was tested in the full-scale tunnel at Langley in March 1945, at the request of the Navy's

The large 1/7-scale powered Kaiser Cargo Wing model undergoing an aerodynamic test in the full-scale tunnel with power on. Note that the model is in its modified configuration with the four vertical tail fins. The proposed cockpit canopy is also noticeable between the nacelles on the upper wing surface. (Photo courtesy of NASA)

Rear overhead view of the huge Kaiser Cargo Wing model mounted for aerodynamic, stability, and control testing on December 18, 1946. The model wing almost spanned the 60-foot-wide entrance cone of the tunnel test section. (Photo courtesy of NASA)

Bureau of Aeronautics. The tests were designed to assess the aerodynamic characteristics of the airplane and to predict its stability and control qualities. During the tunnel tests, the model was tested at various power-on conditions, and the effects of elevator, rudder, and aileron deflections were studied as well as the stalling characteristics and wing profile drag (assessed through wake profile survey analysis). Pitching-moment studies indicated that the model was longitudinally, laterally, and directionally stable for most conditions, but the model was longitudinally unstable at high angles of attack while the propellers were in a windmilling condition. The tests also showed that, as expected, the rudder effectiveness and directional stability of the model were much lower than similar parameters for conventional aircraft. A few months later, the model was modified to permit a comparison of tailless and twin-boom tailed versions of the design. The model was modified by inverting the wing (the original tailless design had wing reflex at the trailing edge sections), removing the vertical tails, and installing a twin-boom tailed configuration. Tested in late 1945 and early 1946 for the same power and attitude conditions in the tunnel, the twin-boom version of the model exhibited aerodynamic characteristics generally similar to the tailless design.

The Kaiser Cargo Wing concept eventually lost out to the famous "Spruce Goose" giant flying boat design that was developed by Howard Hughes in cooperation with Kaiser and test flown in 1947. No Kaiser Cargo Wing aircraft were ever produced, but the design certainly qualified as a radical airplane concept.

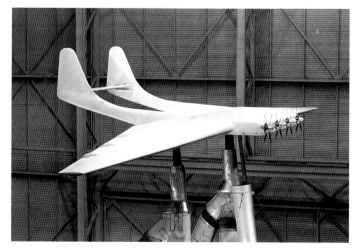

Side view of the twin-boom tailed version of the Kaiser Cargo Wing model during tests. Characteristics were generally similar to the tailless design. (Photo courtesy of NASA)

The first XFG-1 swept-forward wing glider in towed flight. The aircraft had admirable handling qualities during early flight evaluations, but was later destroyed in a spin accident. (Photo courtesy of National Museum of the USAF via Charles L. Day)

Cornelius XFG-1

George W. Cornelius of Dayton, Ohio, was an advocate of remarkably innovative designs such as freely pivoted wings and forward-swept wing configurations during the early 1940s. Under his leadership, the Cornelius Aircraft Co. received a contract in October 1943 to develop an unconventional fuel-carrying glider for the Army Air Forces that featured a tailless forward-swept wing design.

The concept of an unpowered, towed fuel glider had been advanced as a possible application for extending the range of the towing aircraft. The specific operational scenario application was the long-range bomber war in the Pacific. Cornelius's large (54-foot wingspan) glider, known as the XFG-1 (Experimental Fueling Glider) was selected for evaluations, and two experimental aircraft were built by the Spartan Aircraft Corp. at Tulsa, Oklahoma, and delivered to the Army. The wing, vertical tail, and fuselage center section were all of wood construction, while the fuselage nose and tail areas were made of fabric-covered welded-steel tubing. The wing of the XFG-1 was swept forward 15 degrees at the quarter chord, and variable-dihedral fittings were used to provide dihedral angles of three, five, and seven degrees. The fuselage section between the front wing spar and the cockpit contained fuel tanks that held approximately 700 gallons of fuel.

At the request of the Air Technical Service Command, Langley conducted spin-tunnel tests of a 0.056-scale model of the XFG-1 in its 20-foot Spin Tunnel in July 1944. The scope of the testing was to include free-spinning, longitudinal-trim, and tumbling tests for the configuration. Inverted spin behavior and requirements for a spin-recovery parachute were also

Spin-tunnel model of the XFG-1 tested in the Langley 20-Foot Vertical Spin Tunnel. Note deflections of wing spoilers, ailerons, and rudder. Extensive spin and recovery testing took place in the tunnel, including a determination of recommended spin-recovery control procedure for the pilot and the size parachute required for emergency spin recovery. (Photo courtesy of NASA)

studied. The XFG-1 used a conventional vertical tail surface, but it had no horizontal tail surface, the elevator controls being on the wing's trailing edge inboard of the ailerons. The airplane used wing spoilers for precision control in landing. Two models were built for the NACA tests and used in the spin tunnel and the free-flight tunnel.

The results of the spin-tunnel test disclosed that the spin-recovery characteristics of the configuration were quite complicated. The motion of the model in a spin was very oscillatory and disorienting in pitch, roll, and yaw. Although the spin could be terminated by reversing the rudder, the elevator had to be fully deflected in a nose-down direction to effect recovery from an ensuing deep stall. At the same time, too much nose-down control would result in an inverted spin. Although the XFG-1 model would tumble, the tumbling motion could be stopped by deflecting the elevators against the rotation. Researchers recommended that a 7.5-foot diameter parachute with a 27-foot towline would produce satisfactory spin recoveries by parachute action alone for the airplane flight tests.

In April 1945, Langley also conducted free-flight wind-tunnel tests for the Army Air Forces, using the second NACA model of the XFG-1 to determine its stability during towed flight. The specific objective was to determine if a trifurcated towline arrangement would be satisfactory for a large 0.25-scale radio-controlled model of the XFG-1, which the Army planned to spin test at Wright Field. It was intended to tow the 0.25-scale

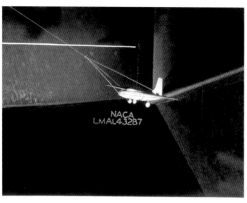

Free-flight tests of the trifurcated towline for the XFG-1 model in Langley Free-Flight Tunnel. Results were used to help guide planned flight tests of a large model at Wright Field. (Photo courtesy of NASA)

model to an altitude of about 4,000 feet, where it would be released, put through a predetermined set of spin-evaluation tests, and then lowered to the ground by an auxiliary parachute.

The NACA free-flight tests were carried out using the trifurcated towline and involved simulations of flights with standard and unconventional loads, variations of the center of gravity, and the use of spoilers. The dynamic stability of the model was determined through testing across a broad spectrum of airspeeds, various control-surface settings, and analysis of glide-path angles. In the tests, it was discovered that no stable towed flights could be maintained with the wing spoilers retracted. During such tests the model was apparently unstable, giving the appearance of riding on a ridge and falling off on either side. Stable towed flights were maintained with spoilers deflected and with the airspeed, longitudinal trim, and glide-path angle set in a combination just before a stalling condition was reached. Langley researchers also discovered that variations of the center of gravity had little impact on the overall stability of the design. Elevator deflections led to more stable towed flights over a broad speed range.

Initial Army flight-test evaluations of the XFG-1 were favorable, and the flight-test program accelerated with additional studies of the aircraft's handling qual-

ities. Unfortunately, on March 14, 1945, the aircraft was destroyed and the pilot killed in an accident during spin tests. Motion-picture records and examination of the wreckage indicated that the pilot did not use the NACA-recommended spin-recovery controls, nor did he deploy the emergency spin-recovery parachute fitted to the aircraft. Flight testing of the second XFG-1 was then terminated until spin studies and recommendations could be obtained from the NACA Langley Spin Tunnel and the 0.25-scale flying model at Wright Field.

Subsequently, a second entry in the spin tunnel, in mid-1945, was conducted to assist in the analysis of the accident and provide data for planned fight tests of the second aircraft. The testing at Langley actually began after the Army flight tests had already begun. Analysis of the Langley staff and the Army engineers at Wright Field included the magnitude of elevator and rudder forces required to recover from spins and post-stall motions.

Following the Langley studies, additional flight tests of the second full-scale XFG-1 continued. However, another non-fatal delayed-spin-recovery incident occurred, causing additional Army concern over the operational viability of the concept. The planned 0.25-scale XFG-1 model at Wright Field was completed and successfully tested by launching from a blimp at the Navy's Lakehurst Naval Air Station, New Jersey. The model test program included deployment of the emergency spin-recovery parachute and recovery of the model after tests by an auxiliary parachute. The conclusion of World War II resulted in the termination of the XFG-1 program.

P-80A Tow Tests

As a result of its participation in numerous military aircraft development programs and the development of extensive expertise in towing aircraft, Langley was frequently called upon for evaluations of radical concepts. As the United States developed its early jet fighters in the mid-1940s, it became apparent that the relatively short range of these aircraft would hinder their versatility in air operations. In 1944, the Air Technical Service Command requested that Langley conduct investigations of the dynamic stability of a towed model of the Lockheed   P-80A Shooting Star to obtain a stable tow arrangement. The objective was to reduce the risk in developing a towing arrangement in which the P-80A would be towed by a B-29 in order to provide fighter protection for bombers over the target area on Pacific

Free-flight model of the Lockheed P-80 Shooting Star used in the towing experiments in the free-flight tunnel. The tow bar located in the nose section of the model is not visible in the photograph. (Photo courtesy of NASA)

missions beyond the effective range of the jet fighters. When fighter protection was required, the P-80A would release from its tug and protect the bombers until the danger of attack was over. For the return trip, the fighter would again hook up with its bomber tug and be towed back to its home base. In order for the tow system to be approved, it was required that complete control-fixed stability be achieved under tow for the cruising condition (180 to 200 mph); that there be only one attachment point and that it be visible to the pilot of the fighter; and that the towline not exceed a full-scale length of 40 feet.

The Langley tests of a 0.059-scale P-80A model in 1944 were conducted for three different tow systems. The first system was proposed by the Army, consisting of a short tow bar mounted in a nose of the P-80A and connected to the tug via flexible towline. A tow hook mounted at the end of a short tow bar on the fighter made contact with a pickup hook mounted on the end of a strut lowered from the tug. The pickup hook was attached to a 40-foot long flexible towline that ran through the strut to a cable drum in the bomber tug. After hook-up, the towline was allowed to reel out and the strut was freed in pitch and yaw.

The second tow-system candidate was conceived by the NACA, consisting of a longer tow bar between the P-80A and the tug with no flexible towline. The fighter would be attached to the bomber's pickup hook either rigidly or with freedom and pitch at the lower end of the tow bar.

The third system investigated was a trifurcated bridal system for the fighter, consisting of a nose line and two wing lines from the fighter that joined at the apex where the towline was attached. Although it did not meet the requirement of a single attachment point, the NACA tested the configuration because it might have proven to be a satisfactory alternate arrangement.

Results of the tests concluded that the Army's original short tow-bar arrangement failed to produce stable tows with the proposed towline attachment point. Unstable lateral oscillations occurred with almost constant amplitude. Attachment points behind and above the proposed specific tow-bar location on the fighter model resulted in more satisfactory characteristics, but did not result in stable tows. With the long tow-bar arrangement, stable tows were possible for some attachment points and stability was noticeably better than the original proposal. The stability characteristics of the model with the trifurcated tow arrangement were similar to those of the long tow-bar arrangement.

In September 1947 Wright Field conducted exploratory flight testing of a modified P-80A behind a B-29 tow airplane. For these tests the Army chose the tow-cable approach. Following several attempts to link with the B-29, the pilot finally succeeded and was towed for about 10 minutes, but he experienced great difficulty in releasing the tow bar. The event caused enough concern to result in cancellation of the project. However, the subject of towed fighters for strategic protection remained a topic of interest in postwar years, as will be discussed in the next chapter.

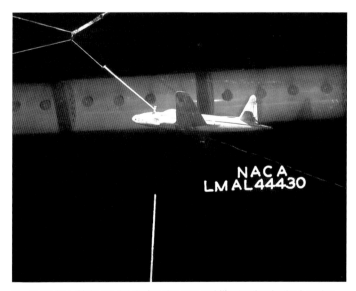

The P-80 free-flight model during the tow stability testing in the Langley Free-Flight Tunnel in 1944. The tow configuration initially proposed by the Army Air Forces was found to be unsuitable, and flight tests were conducted with a towline. (Photo courtesy of NASA)

CHAPTER 3

POSTWAR INNOVATION

After World War II, the aircraft industry in America was experiencing a boom in revolutionary airplane designs. New, sleek, jet-powered designs gradually replaced classic propeller-driven aircraft. Radical aeronautical concepts, some of which were only on the drawing board during the war, now found their way to the runway and were evaluated in the test facilities of the NACA. Military and aviation planners were quick to pursue the development of some of these concepts to meet a new threat emerging from Communist Russia. The growing threat of the Soviet Union and the absorption of eastern European states into what became known as the Warsaw Pact signaled the beginning of the Cold War. This struggle between democracies of the West and Communist states of the East lasted nearly a half century and gave rise to numerous generations of aeronautical technologies, including countless unconventional aircraft designs, many of which were never mass produced.

At Langley, research methods that were only in their infancy during World War II became mainstay techniques in the analysis and study of leading-edge aircraft designs. These research methods provided engineering data for further development and maturation of new

aeronautical concepts. The full-scale tunnel continued to play a vital role in testing large-scale models of radical aircraft designs, and began to support research performed with powered and unpowered free-flying tethered models. Stability and control studies performed in the free-flight wind tunnel contributed to the testing of postwar radical aircraft designs, and the spin tunnel staff was swamped by additional studies to determine the effects of new geometric variables such as swept wings. The impact of jet propulsion and supersonic flight brought another round of intense work to the center.

Lippisch DM-1

Toward the end of World War II, Germany was planning flight tests for a revolutionary new aircraft design to serve as an interceptor against the overwhelming Allied heavy bomber formations that were pounding the Third Reich. Legendary German aerodynamicist Alexander Lippisch had begun experimenting with tailless and delta-winged aircraft designs during the 1920s and throughout the 1930s. By the time World War II broke out, his services had been enlisted in the development of the subsonic Messerschmitt Me 163

Komet rocket-powered interceptor, and he continued work on a revolutionary delta-wing design that was to serve as a supersonic ramjet interceptor, the Lippisch P13a. A full-size version of the P13a—the wooden DM-1 glider—was under construction in Germany to investigate the design's low-speed, high angle-of-attack characteristics when the war ended.

Under the auspices of Operation Lusty, the acquisition of secret Luftwaffe technologies, the DM-1 glider and other advanced German aircraft were confiscated by Allied technical intelligence personnel. The construction of the glider was completed in Germany, and it was transported via ship to NACA Langley for aerodynamic testing in the Langley Full Scale Tunnel in February 1946. The primary objective for the test was to obtain full-scale data on the maximum lift and stalling characteristics of highly swept delta wings for correlation with small model results.

The initial test results of the DM-1 glider in the full-scale tunnel indicated that the configuration's maximum lift coefficient was considerably lower than had been indicated by previous small-scale model tests in Germany and several U.S. wind tunnels. Maximum lift coefficient measured for the original DM-1 glider was only 0.6, which was about 0.3 less than maximum lift coefficient measured for sub-scale models of delta wings having about the same aspect ratio. In addition, the DM-1 wing stalled at an angle of attack of about 18 degrees, whereas the sub-scale models typically did not stall until angles of attack reached about 40 degrees. As a result of the poor correlation, the NACA test program was focused on understanding why the lift was so low, and how to improve the capabilities of the DM-1.

Several modifications were made to the glider during the full-scale tunnel tests. The aircraft had been designed with an airfoil section similar to the NACA 0015-64 section, with a relatively bulbous nose shape. As received, the glider was equipped with a rudder for directional control and elevons for lateral and longitudinal control. However, the balance gaps on the control surfaces were relatively large. Researchers modified the glider by adding a sharp leading edge to the wing semispan, removing the vertical fin, and sealing the

The Lippisch DM-1 delta wing glider at an airfield in southern Germany following its capture by American forces in May 1945. The DM-1 glider was later brought to America for aerodynamic testing in the Langley Full Scale Tunnel under Operation Lusty, which directed the seizure of examples of advanced German technology for U.S. evaluations. (Photo courtesy of NASA)

The Lippisch DM-1 glider mounted in the full-scale tunnel for aerodynamic testing. Test program focused on the high-lift, low-speed aerodynamic characteristics of the delta-wing configuration. The maximum lift of the DM-1 was found to be much lower than values obtained for generic sub-scale models in the United States, leading to an early study of the control of vortex flows. (Photo courtesy of NASA)

DM-1 in its original configuration during testing in the full-scale tunnel on April 16, 1946. Note the wool tufts on the upper surfaces of the aircraft used for visualization of the airflow properties. (Photo courtesy of NASA)

control-balance slots. These full-scale tests were augmented by additional Langley tunnel tests of small delta-wing models having either thick or thin wing sections.

The full-scale tunnel tests of the DM-1 configuration were milestone events for Langley's aerodynamic experts. The highly swept sub-scale delta-wing models exhibited evidence of vortex flows on the upper surface at high angles of attack near maximum lift conditions; however, the full-scale DM-1 wing did not. This fundamental difference in aerodynamic behavior was attributed to the fact that the large leading-edge radius of the DM-1 wing suppressed the formation of vortex flows and suggested that a sharp leading edge could fix flow separation at the leading edge and create powerful vortical flows for lift augmentation. When a sharp leading edge that extended halfway across the span was added to the DM-1 wing, the maximum lift coefficient was dramatically increased to a value of 1.01 at an angle of attack of 31 degrees. Additional modifications of removing the vertical fin and sealing the large control-balance gaps increased the maximum lift coefficient to 1.24.

The major conclusion of the DM-1 test program was that airfoil sections having sharp leading edges or small leading-edge radii considered desirable for supersonic flight might also have acceptable low-speed high-lift characteristics for highly swept delta aircraft. It was noted, however, that the angles of attack required to produce high lift at low speeds for delta-wing configurations would be considerably greater than those for conventional aircraft. This early Langley recognition

Side view of yet another Langley research modification to the DM-1. Note the bubble-type canopy, revised vertical tail, and modified wing with beveled sharp leading edges. Research performed using the DM-1 glider produced data and design guidelines that assisted industry in the development of America's delta-wing fighters, including the XF-92, F-102 Delta Dagger, and F-106 Delta Dart. (Photo courtesy of NASA)

and use of vortex flow to increase lift at high angles of attack was an important precursor to the current practice of generating vortex lift with sharp-edged auxiliary lifting surfaces. Examples of today's aircraft using vortex lift include the F-16 and F/A-18.

Results obtained from the analysis of the aerodynamics of the DM-1 glider served as the basis for the American XP-92 that was slated for service as a short-range interceptor by the Air Force. Although the XP-92 program was eventually cancelled, the basic delta-wing concept was later used in the experimental XF-92 program which led directly to the development of the highly successful supersonic Convair F-102 Delta Dagger and Convair F-106 Delta Dart designs that saw service with the U.S. Air Force from the 1950s to the 1990s.

Oblique Wing

One of the most unconventional aeronautical concepts explored in the post-World War II-era was the oblique or "switchblade" wing concept. This concept had its roots in Germany during World War II when Richard Vogt of the Blohm & Voss Flugzeugwerke designed an oblique-wing twin-engine jet fighter design, known as the Blohm & Voss P 202, in 1942. The design was part of the German thrust of using wing sweep to delay the onset of transonic drag increases. The aircraft design was to feature a wing that rotated in flight and was positioned on the upper surface of the fuselage. The Blohm & Voss P 202 never advanced past the drawing-board stage, but the Messerschmitt Flugzeugwerke later proposed several other advanced oblique-wing jet-fighter designs before the end of the war. However, these designs were never pursued either.

At Langley, Robert T. Jones led an aggressive program that had independently identified the beneficial effects of wing sweep on delaying compressibility effects. However, the use of wing sweep introduced several undesirable aerodynamic problems at the low speeds used for takeoff and landing. For example, tip stall was prevalent on high-aspect-ratio swept wings, resulting in pitch up; the effective dihedral was very high; and the drag at moderate angles of attack was also high. To gain the advantages of sweep at high speeds and avoid the low-speed problems, researchers proposed pivoting the wing so that it could be set in the conventional unswept position for takeoff and landing, and then swept back for high-speed flight.

In 1946, Langley began to explore the potential of the oblique-wing concept when a team of researchers led by John P. Campbell, working under the direction of Jones, conducted wind-tunnel tests of an oblique-wing model in the Langley Free-Flight Tunnel. These studies were aimed at investigating the dynamic stability and control characteristics of a skewed-wing aircraft design with its asymmetric shape. Initially, many in the aeronautical community expressed great skepticism, unable to envision the feasibility of attempting to control such a radical design. During the Langley tests, free-flight evaluations, aerodynamic force and moment measurements, and other aerodynamic analyses were performed with the model. The project was intended to be totally exploratory, without detailed analysis of whether such a radical aircraft design would be feasible. In keeping with this philosophy, the model used for the tests was a rudimentary boom-wing-tail arrangement. The rotatable wing was pivoted at the 50-percent-chord position and could be skewed up to an angle of 60 degrees.

The Langley researchers concluded from the free-flight tunnel tests that stability and control could be adequately maintained with the wing skew angle between 0- and 40-degrees. Longitudinal stability and control were satisfactory, and only a small longitudinal trim change was noted as the wing was skewed to 40 degrees. However, when the wing skew angle of the model was set to 60 degrees, lateral trim changes increased and the aileron control was inadequate. The Langley free-flight tunnel tests of the oblique wing model provided useful initial data for future American oblique-wing research. It was later determined by researchers that wing skew angles in excess of 40 degrees would be required for supersonic aircraft and that larger models needed to be studied. R. T. Jones subsequently transferred from Langley to the NACA Ames Aeronautical Laboratory, located in Mountain View, California, but maintained his interest in the oblique-wing design.

The oblique-wing concept was revisited during the 1970s in NASA research efforts at Ames. Jones and his team of engineers studied and proposed the idea of an oblique-wing supersonic transport. The design was examined in wind tunnels, but the idea was terminated when plans for an American supersonic transport were scrapped during the early 1970s.

A low-speed oblique-wing demonstrator aircraft design was later conceived and tested in wind tunnels at Ames and Langley. The design, designated the Ames-Dryden-1 or AD-1, was made of advanced composite materials and intended for low-speed flight testing. The low-cost demonstrator was built by the Ames Industrial Co. of Bohemia, New York. Langley supported the AD-1 flight program with spin-tunnel tests of a 0.077-scale model in 1981. Interestingly, the spin-tunnel results showed that the AD-1 model would not spin in the direction of the swept-back wing panel. That is, no spins could be obtained to the left when the wing was skewed with the right wing swept forward.

Flight tests of the radical AD-1 design began at the NASA Dryden Flight Research Center in 1979. Testing

Oblique-wing free-flight model undergoing dynamic stability and control flight testing in the Langley Free-Flight Tunnel in 1946. The model's pivoted wing has been set at a 40 degree skew angle. Although handling qualities were satisfactory for skew angles up to 40 degrees, when sweep was increased to 60 degrees major lateral-control problems developed. (Photo courtesy of NASA)

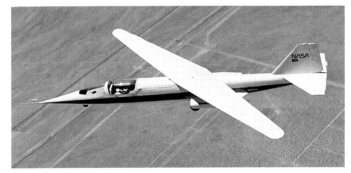

By the late 1970s, oblique-wing research had reached maturation and a full-scale manned aircraft known as the AD-1 was test flown at NASA Dryden Flight Research Center as a low-cost concept demonstrator. In this photo, taken on January 1, 1980, the AD-1 effectively demonstrated a 60-degree wing-skew angle in a test flight over NASA Dryden. (Photo courtesy of NASA)

continued through 1980, with the aircraft successfully demonstrating a wing skew angle of 60 degrees in stable flight at a speed of 150 knots. Its last flight in the 79-flight program occurred on August 7, 1982. The flying characteristics of the AD-1 were, in general, similar to the earlier Langley free-flight model in that the handling qualities for wing sweep angles above about 45 degrees were poor due to pitch-roll coupling effects.

Bell L-39

The aerodynamic benefits of swept wings to delay compressibility drag effects quickly became of interest to the U. S. military, and the services moved aggressively to identify and solve any problems associated with this revolutionary concept. The Air Force concentrated on the high-speed portion of the flight envelope for its swept-wing studies, but the Navy focused on the low-speed envelope and the potential application of swept-wing aircraft for carrier-based operations.

The swept wing was known to exhibit poor low-speed characteristics created by the flow of the boundary layer toward the wingtips, resulting in tip stall that adversely affected stability and control. One of the Navy's first flight projects to investigate the handling qualities and low-speed aerodynamic characteristics of swept-wings was initiated with a 1945 contract to the Bell Aircraft Corp. to modify two conventional Bell P-63 Kingcobra aircraft with swept wings for flight-evaluation tests. The modified aircraft, which were designated L-39-1 and L-39-2, were fitted with outer-wing panels swept back 35 degrees at the wing quarter-chord location, and with a short unswept section on the inner wing inboard of the air scoops. The wings were designed specifically to test the swept-wing configuration with different bolt-on leading edges. The L-39-1 aircraft was built with a NACA low-drag wing airfoil section, while L-39-2 was built with a bi-convex, circular-arc wing section. The first research aircraft flew at Bell in April 1946, and after a four-foot extension and a ventral fin were added to the rear fuselage to improve directional stability, Bell conducted flight tests to evaluate the handling characteristics of various leading-edge devices. The L-39-1 aircraft was then assigned to NACA Langley in August 1946 for detailed aerodynamic and handling-quality investigations.

Prior to the arrival of L-39-1, Langley researchers had conducted preliminary wind-tunnel tests of the configuration in preparation for the flight-test program. Spin-tunnel tests were conducted, and data were obtained from tests of a 0.22-scale L-39 model in the Langley 300-mph 7-by-10-Foot Tunnel in May 1946. The tests in the 7-by-10-foot tunnel included investigations of the stability, control, and stalling characteristics of the model with several leading-edge shapes including plain, slotted, and drooped-nose configurations. A circular-arc wing airfoil section was also evaluated. The wing of the wind-tunnel model permitted various configurations of leading-edge slots to be adapted including

Model of the Bell L-39 research aircraft mounted for tests in the Langley 300 mph 7 x 10-Foot Tunnel in 1946. NACA flight data obtained from flight tests of the full-scale L-39 at Langley were correlated with the data derived from the wind-tunnel tests. Note the leading-edge slot configuration and the gear-down configuration. The main gear was fixed in the extended position for L-39-1. (Photo courtesy of NASA)

Rear view of the L-39 model shows the leading-edge slot configuration to advantage. Note the 3-blade propeller used by the L-39, rather than the 4-blade configuration used by conventional P-63 aircraft. (Photo courtesy of NASA)

40-, 60-, 80-, and 100-percent span lengths. The tunnel tests were conducted for the model with the main wheels down, since retraction of the main wheels on the L-39 full-scale aircraft was not possible (the nose gear, however, could be retracted).

Results of the tunnel tests indicated that, as expected, the variation of lift with angle of attack for the slot-equipped wing configuration was less than that of the plain wing, but that the angle of attack for maximum lift was increased about 5 degrees. The effectiveness of the leading-edge slots was directly related to the span-wise extent of the slot configuration. The slots also greatly increased the lateral-directional stability of the configuration. When configured with the circular-arc leading edge (resulting in a sharp leading-edge), the model exhibited large variations of longitudinal stability with angle of attack. Although stable at low angles of attack, the model became unstable at moderate angles of attack. Researchers also found that although the directional stability of the configuration was increased at low angles of attack by lengthening the rear fuselage, the addition of a ventral fin was required to obtain satisfactory stability at high angles of attack.

Bell L-39-1 in flight. During flight-research studies at Langley, NACA pilots found that the design possessed good directional stability throughout the speed regime evaluated. Data from the earlier Langley wind-tunnel tests correlated well with the data derived from the flight studies. Although the L-39 was not intended to become a production design, information obtained from the L-39 program directly benefited the development of the Navy Douglas D-558 Phase II supersonic research aircraft. (Photo courtesy of NASA)

Flight investigations of the L-39-1 aircraft by the NACA were carried out at Langley to study the low-speed lateral-directional stability and control characteristics of the swept-wing configuration. The airplane was flown without slots on the wing, and with slots along 40- and 80-percent of the span. Flights were also made without the large ventral-fin extension of the normal flight-test configuration.

During the L-39 flight research studies, Langley pilots found that the aircraft exhibited good directional stability throughout the speed regime evaluated. In most categories, the wind-tunnel data for the configuration with the 80-percent-span slots was in excellent agreement with the data derived from the flight studies, including dihedral effect and aileron rolling effectiveness. When the large ventral fin was removed, researchers observed a dramatic decrease in directional stability at high angles of attack, which was also in agreement with the preceding wind-tunnel test results. At one point in the flight-test program, the wing leading edge of L-39-1 was modified with a circular-arc section; however, the flying qualities of the modified aircraft were poor, and the modified configuration was judged to be unsuitable for further tests.

Flight testing of the L-39-1 continued at Langley through June 1949, with a total of 101 test flights having been made. The second aircraft, L-39-2, conducted all its flight tests at the Bell site with particular interest in the circular-arc airfoil to obtain data for the Bell X-2 supersonic test airplane which was under design. The L-39-2 was subsequently transferred to Langley in 1949, but it was not flown by the NACA. Both aircraft were later shipped to the NACA Lewis Laboratory, where the NACA staff used the aircraft in research on crash and fire-suppression technology.

The L-39 aircraft were intended to be research tools and flight demonstrators without any perspective of production in mind. As such, their most important contribution was a demonstration of satisfactory low-speed, high angle-of-attack behavior for swept-wing aircraft. Pilots were very pleased with the baseline L-39-1's behavior during stalls with no wing leading-edge devices. When the aircraft was modified with leading-edge slots it became even more docile during stalls, without wing-dropping behavior or excessive loss of altitude. These results clearly demonstrated to the Navy that it might be possible to employ swept-wing aircraft satisfactorily for carrier-based operations.

McDonnell XP-85 Goblin

After World War II the problem of providing fighter protection for bombers on long-range bombing missions became a serious issue. As the Cold War with the Soviet Union began to intensify, military planners envisioned the United States having a large fleet of long-range heavy nuclear bombers with stowed or attached "parasite" fighter aircraft that bombers could deploy in case they were threatened by enemy interceptors. In early 1945, the Army Air Force requested proposals from industry for a parasite fighter that could be carried within the Convair B-36 bomber. The winning design by the McDonnell Aircraft Corp. was the XP-85 Goblin, and two prototypes were ordered in October 1945.

The mother B-36 would deploy and retrieve the Goblin from its bomb bay for self-protection. The little jet-powered Goblin was to be armed with four 0.50-cal. machine guns. The Goblin was designed to be lowered from the B-36 on the trapeze, launched for combat, and hooked back onto the trapeze upon return from its mission. The aircraft would then be secured, its wings would be folded (resulting in a folded span of five feet), and then it would be drawn up into the bomb bay. In the event the Goblin was forced to make a landing, the aircraft was equipped with a steel skid beneath the fuselage.

In 1947, the Air Materiel Command requested that Langley perform free-flight tunnel studies of the low-speed, power-off, stability and control characteristics of the XP-85 configuration. The tests consisted of two phases. First, aerodynamic characteristics of a 0.20-scale model of the XP-85 were measured in conventional force tests, then a study was undertaken to determine the dynamic stability characteristics of another 0.10-scale XP-85 model while hooked onto the trapeze and during retraction into a simulated B-36 bomb bay. For the retraction tests, a 0.10-scale model of the forward bomb bay of the B-36 was fabricated and attached to the ceiling of the free-flight tunnel. A scale model of the full-scale trapeze was attached to the bomb bay and used as a support for the model. A nose-restraining "collar" attachment was used to clamp the model in position for retraction into the bomb bay. Using this approach, the Langley researchers examined the stability problems associated with the hook-on maneuver to the trapeze, folding of the XP-85's wings, and retraction into the bomb bay.

The McDonnell XP-85 Goblin was a radical parasite fighter designed to be carried, launched, and recovered from the bomb bay of the B-36 bomber. The little fighter is shown here on the tarmac at St. Louis. Note the multiple vertical tail surfaces required for directional stability of the stubby configuration. The Navy had pioneered the parasite concept for the U.S. military in the 1930s when Curtiss F9C-2 biplanes were designed for carriage on the airships Macon and Akron. (Photo courtesy of The Boeing Company)

View of the XP-85 attached to the bomber trapeze apparatus shows details of the concept. The folded wings of the XP-85 allowed packaging within the B-36 bomb bay during carriage or retraction. Note the engagement hook of the XP-85 on the trapeze member, and the nose-restraining collar used to restrain the aircraft on the trapeze. (Photo courtesy of the National Museum of the Air Force)

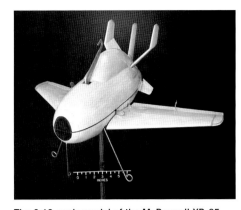

The 0.10-scale model of the McDonnell XP-85 Goblin used for dynamic retraction tests in the Langley Free-Flight Tunnel. The study included an analysis of the model's stability while hooked onto the "parasite" trapeze and during retraction into a simulated B-36 bomb bay. The model was also tested with folded wings on the trapeze. (Photo courtesy of NASA)

The XF-85 on display at the National Museum of the U.S. Air Force at Wright Patterson Air Force Base, Ohio. During flight tests the XF-85 exhibited low levels of directional stability, and vertical fins were added to the outer wings for additional stability. View also shows details of the engagement hook. (Photo courtesy of the National Museum of the Air Force)

Results of these studies showed that stability of the XP-85 aircraft design was adequate for power-off conditions and that the special attachment collar could be easily positioned. When tests were conducted in a simulated power-on condition by increasing the tilt of the free-flight tunnel (increasing the angle of attack of the XP-85), the model displayed undesirable constant-amplitude rolling and sidewise motions, particularly if the collar made contact with the model without making adequate attachment. In one instance, the collar contacted the model but missed the nose. The ensuing model motions in roll and yaw exceeded 90 degrees, and would have been catastrophic at full-scale conditions. If the model was locked into the appropriate position by means of the collar, it was determined that the wings could be properly folded and the aircraft could be easily stored in the bomb bay.

Concerned over the dynamics of the model on the trapeze, Langley engineers suggested that power be turned off once the Goblin hooked onto the trapeze and that a "restricting mechanism" in the mid-section of the trapeze be added to prevent undesirable sidewise motions. Researchers also suggested that the XP-85 pilot or spotter in the bomber aircraft direct the retraction effort for better safety and ease of operation.

Two Goblin test aircraft were produced, and the first flight tests occurred in August 1948, after the first aircraft was tested in the NACA 40- by 80-Foot Tunnel at the NACA Ames Aeronautical Laboratory. The B-36 was not yet available for the flight tests, so an EB-29B bomber was used as a substitute. On the first flight, McDonnell test pilot Ed Schoch launched from the EB-29 successfully over Muroc (now Edwards AFB), but during the attempt to hook up at the end of the flight he missed the trapeze and shattered the canopy of the Goblin by hitting the trapeze structure. Schoch was able to land the damaged XP-85 (by then redesignated as the XF-85) on the desert. During the seven-flight test program, the Goblin successfully connected to the EB-29B trapeze on only three flights due to the effects of turbulence near the bomb bay. No flight tests were made using a B-36 as the mother ship. The XF-85 parasite concept was abandoned in late 1949 in favor of evolving aerial-refueling concepts for fighter escorts. Both XF-85 aircraft survived, with one on display at the National Museum of the U.S. Air Force at Wright-Patterson Air Force Base and the other at the Strategic Air and Space Museum at Ashland, Nebraska.

Fighter/Bomber Wingtip Coupling

The U.S. Air Force also explored another parasite concept known as fighter/bomber wingtip coupling during the early 1950s. The studies had, in part, been stimulated by a concept advanced by German immigrant Dr. Richard Vogt to the U.S. Air Force at Wright Field after

World War II, based on limited flight testing that he had directed in Germany during the war. Vogt had been part of the group of scientists brought to the United States as part of Operation Paperclip. Vogt's concept consisted of coupling a parasite fighter to each wingtip of a bomber and deploying the fighters when protection was required. By hinging the fighters to the bomber's wingtips with freedom in roll, the wing bending loads caused by the fighters' aerodynamic and mass forces could be minimized. A series of feasibility tests were conducted by the Air Force beginning in 1949 using a C-47 and a Q-14 Culver Cadet. More than 230 couplings were successfully flown in the program, and a vast amount of experience was gained for future designs. In 1950, a follow-on project nicknamed "Tip Tow" was initiated by Republic Aviation in which two F-84 Thunderjets successfully maintained wingtip coupling with a B-29 bomber.

The concept of fighter/bomber coupling was studied extensively in research projects by NACA in the Langley Free-Flight Tunnel. In 1949, studies were conducted in the tunnel to explore the flying characteristics of a generic model having self-supporting models of fuel-carrying panels hinged to the wingtips. In theory, the panels increased the aspect ratio of the basic aircraft wing, resulting in reduced drag for extended range.

In 1951, Langley researchers also conducted a study of the dynamic longitudinal stability and control of a generic fighter/bomber model configuration with rigid and pitch-free couplings. The study involved simulated flight tests of a "freely coupled" model combination, with the models

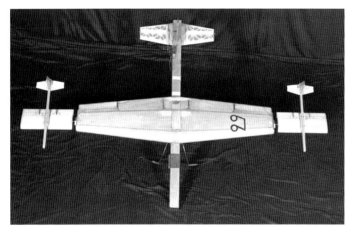

Researchers in the Langley Free-Flight Tunnel conducted generic research on coupled aircraft configurations to evaluate concepts to extend the range and increase the protection of bomber aircraft. This model was used in 1949 to study the dynamic stability characteristics of a carrier aircraft with flying models of fuel tanks attached to each wingtip. (Photo courtesy of NASA)

being joined at the tail of the bomber and nose of the fighter, having freedom in pitch. Simulated flight tests of the bomber model by itself were also performed. Results of the tests indicated that stability and control of the pitch-coupled configuration were adequate and even better than for the bomber model by itself. However, the test results also indicated that the rigidly coupled configuration was woefully unstable. Langley engineers concluded that the instability of the rigid condition was caused by the downwash of the bomber model on the fighter model. Consequently, this type of coupling was never attempted with actual aircraft in flight.

In 1949, Langley aerodynamicists conducted a theoretical investigation of the aerodynamic benefits of wingtip coupling of a B-36 and two B-47 bombers. Three years later, in 1952, free-flight tunnel researchers explored the concept of fighter/bomber wingtip coupling in several wind-tunnel studies using simple fighter and bomber models. In one investigation, a rudimentary B-36-type bomber model and two swept-wing fighter models were used in free-flight tests. The tests were conducted to explore the dynamic lateral stability and control characteristics of the wingtip-coupled configuration. In the free-flight test setup, the outboard aileron of each fighter was automatically deflected in response to the relative bank angle between the bomber and the fighter to keep the fighter aligned with the bomber wing. For example, as an individual fighter rotated up at its wingtip location, the outboard aileron was deflected upward to reduce lift on that wing panel, and the fighter model tended to return to its original position relative to the bomber wingtip. The investigation consisted of flight tests of the bomber model alone and with the fighters attached with freedom in roll with respect to the bomber. Test variables included the longitudinal position of the fighters with respect to the bomber and the geometry of the ailerons of the fighter models. Test results defined the level of roll control required for satisfactory behavior of the coupled configuration, and it was also found that the flight characteristics of the combination were directly affected by the relative longitudinal location of the fighters and the bomber. The configuration's flight characteristics were found to be best when the fighters were located further to the rear (behind the 0.50-tip-chord) of the bomber.

After many successful launches and retrievals of the fighter parasites, the Air Force's wingtip coupling experiments using F-84s and a B-29 came to a violent end on April 24, 1953. A fatal automatic-control-induced accident

Langley also studied other modes of coupling aircraft. This general research model was used in 1951 to study the characteristics of configurations coupled in pitch. A pivoting hinge behind the lead aircraft was used to couple a second aircraft for an auxiliary fuel source. (Photo courtesy of NASA)

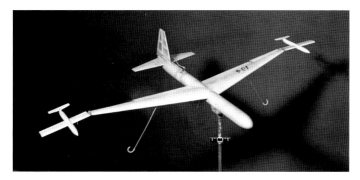

Inspired by theoretical studies at Langley of wingtip-coupled B-36 concepts, researchers at the free-flight tunnel conducted flight studies of this model for general research on stability and control in 1952. The fore-and-aft location of the parasite models was critical to the flight behavior of the coupled configuration. (Photo courtesy of NASA)

An RF-84F fighter is shown coupled to the right wingtip of a B-36 during Project Tom-Tom. Wingtip coupling experiments were later abandoned when in-flight refueling became commonplace. (Photo courtesy of Lockheed Martin via Eric Hehs)

occurred when one of the wingtip-mounted Thunderjets pitched upward, rolled inverted, and then slammed into the wing of the B-29, destroying the bomber's outer wing. The Thunderjet and the B-29, as well as their crews, were lost in the accident.

In 1956, the Air Force once again experimented with the concept of wingtip coupling, this time under a contract with General Dynamics using a B-36 bomber and two swept-wing RF-84F Thunderflash fighters in an experiment that became known as Project Tom-Tom. The Tom-Tom project was supported by flutter testing in the Langley 19-Foot Pressure Tunnel. In these tests, structurally scaled models of the B-36 and coupled RF-84F aircraft were used to determine the various aeroelastic and flutter modes of motion expected in flight.

Several successful in-flight wingtip-coupling combinations were demonstrated during flight, but the program came to an end on September 23, 1956, when one Thunderflash was ripped from the wingtip of the bomber. Despite the loss of a large outer section of the B-36 wing, all crews returned safely. Considering the hazards associated with fighter/bomber wingtip coupling and the emerging practicality of in-flight aerial refueling, the Air Force abandoned the concept of fighter/bomber wingtip coupling following the conclusion of Project Tom-Tom.

Convair XB-53

During World War II, the Germans had progressed in the development of several forward-swept-wing jet designs. These designs included the Junkers Ju 287, which progressed to flight tests, and others that were on the drawing board when the war ended. The United States, under Operation Paperclip, gained access to German expertise with these radical designs, and industry teams began to assess the advantages of the concept. A product of the activities was an American forward-swept-wing tailless jet design for the Air Force that became known as the XA-44, with the A designation denoting "attack." The design incorporated 30 degrees of forward sweep, which offered the advantages of a delay in compressibility effects at high speeds while also retaining good stall and control characteristics at low speeds. Two XA-44 prototypes were to be built by Convair (Consolidated Vultee at the time) and powered by three J35-GE turbojets. In 1946, the XA-44 was renamed the XB-53, with the philosophy that the aircraft would serve as a light bomber.

In 1947, the NACA investigated the aerodynamic and flying characteristics of the XB-53 configuration in the Langley Free-Flight Tunnel. The objective of the test program was to determine the low-speed, power-off stability and control characteristics of a 0.05-scale model of the advanced bomber design. For the high-lift, flap-down configuration, outboard single-slotted flaps were deflected, together with trailing-edge ailerons and inboard wing leading-edge slots.

Prior to the Langley tests, static wind-tunnel force and moment measurements had been made on a 0.10-scale XB-53 model in the GALCIT (Guggenheim Aeronautical Laboratory, California Institute of Technology) tunnel in 1946. Data comparisons between the tunnels indicated good agreement. At high angles of attack, results of both

View of the 0.05-scale model of the Consolidated Vultee (later Convair) XB-53 used in free-flight tunnel studies at Langley. The tests were designed to study the aerodynamic and flying characteristics of the radical forward-swept wing, tailless bomber design. (Photo courtesy of NASA)

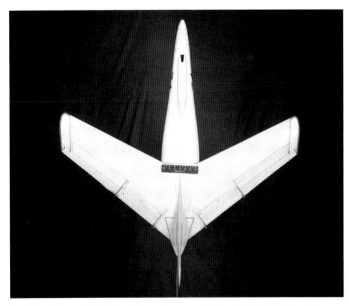

Overhead view of the XB-53 free-flight model showing the 30° swept-forward wing. The results of the free-flight studies were promising, but the XB-53 program was eventually terminated without a prototype being completed. (Photo courtesy of NASA)

tunnel tests showed that the design would exhibit a dramatic reduction in directional stability, attributed to low-energy flow from an area of flow separation on the inner wing impinging on the vertical tail surface. Use of the inboard leading-edge slot devices greatly improved the directional-stability issue.

During the free-flight tests, the most serious problem noted was unsatisfactory response to lateral controls. The outboard wing ailerons produced large values of adverse yaw when roll control was applied, resulting in diminished roll control, large sideslip angles, and wallowing of the model. Increasing the directional stability by adding an extension to the vertical tail was beneficial and delayed the onset of poor stability to higher angles of attack, but this modification did not reduce the poor handling qualities at the stall. The addition of the inboard wing leading-edge slot, together with the increased vertical tail, resulted in satisfactory characteristics. The benefits of the leading-edge slot were apparent for both the free-flight and GALCIT models. The NACA researchers concluded that the combined deflections of the outboard wing flap and the trailing-edge aileron was a major source of flap-down handling-quality issues, and that these difficulties might be remedied by the use of another high-lift concept.

Meanwhile, Convair and the Air Force pursued another design, the XB-46, and work on the XB-53 prototypes was never finished. The radical XB-53 program was terminated in 1949.

McDonnell XF-88B

As the American aircraft industry reached unprecedented growth in the 1950s, the nation's commercial and military aviation planners were confronted with more new advanced propulsion concepts than ever before. Many planners believed that the supersonic propeller—a high-speed propeller married to a turboprop engine—offered better fuel economy than emerging turbojet engines. In this technical atmosphere, the Air Force became interested in a high-speed propeller-driven interceptor and escort fighter for long-range missions. Langley had initiated aggressive wind-tunnel and theoretical research studies on supersonic propellers spearheaded by the famous John P. Stack. The Wright Air Development Center (WADC) of the Air Force also began experiments to determine the feasibility of various supersonic propeller designs with the Hamilton-Standard and Curtiss Wright companies. The WADC and NACA experiments indicated

The McDonnell XF-88B with bulbous spinner and short-blade propeller at Langley in 1956. Note the gages on the propeller blades and the propeller wake survey rake behind the propeller. During flight-research studies at Langley, the XF-88B was outfitted with a variety of different spinner shapes and propeller blades to evaluate aerodynamic efficiency. (Photo courtesy of NASA)

McDonnell XF-88B in flight with its propeller feathered at the McDonnell plant in St. Louis, Missouri, in 1956. The Air Force supplied the airplane and test propellers, and the Navy provided the T-38 turboprop engine for the study. Close inspection of the propeller blades reveals torsion and bending gages used for stress measurements. Three different propeller designs were tested at flight speeds above Mach 1. (Photo courtesy of NASA via Jack Reeder)

that unconventional propeller designs were going to be necessary to offset the well-known abrupt decrease in propeller performance as standard propellers reached a speed of Mach 0.9 and suffered compressibility losses.

To study their supersonic propeller research findings in flight, the WADC devised a flight-test program using a modified Republic F-84F jet fighter. In 1952, Republic began the development of two experimental aircraft that became known as XF-84Hs. These airplanes were outfitted with powerful Allison XT40-A-1 turboprop engines. The single turboprop engine was married to a three-blade propeller enclosed by a large spinner. The first test flights of the XF-84H occurred in 1955 at Edwards Air Force Base, California. The XF-84H program proved to be a major disappointment, as the airplane did not perform well. The aircraft failed to reach supersonic speeds and generated excessive, unbearable noise from its propeller, earning the aircraft the name "Thunderscreech."

At about the same time that the XF-84H was being developed, the McDonnell Aircraft Co. developed another supersonic propeller testbed after being persuaded by the Air Force to pursue such a project. This airplane, designated the XF-88B, was essentially an XF-88A jet fighter outfitted with an Allison XT-38 turboprop. The Air Force, Navy, and NACA formed a partnership to support the XF-88B flight-test program.

In May 1953, test pilot Jack Reeder was detailed to the McDonnell plant in St. Louis to ferry the XF-88B back to Langley. Engineers and flight-research personnel at Langley studied numerous propeller-blade and spinner configurations using extensive instrumentation. Flight tests of the XF-88B indicated that the airplane performed well with the turboprop, easily achieving a speed of Mach

0.95. Three different propeller designs were flown to speeds above Mach 1. The terrible propeller noise levels prevalent in the previous XF-84H also revealed itself in the XF-88B flight-test program. The aircraft did eventually attain supersonic speeds during dives. Unfortunately, the flight tests came at an inopportune time, as interest in supersonic propeller-driven aircraft suddenly subsided. The XF-88B flight-test program was cancelled due to waning interest in the supersonic propeller concept. The competing turbojet-engine concept had proven to be a better performer and was easier to maintain.

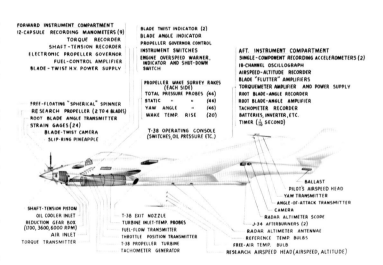

Schematic showing the extensive amount of NACA instrumentation used during flight tests of the XF-88B. Preceded by in-depth NACA wind-tunnel studies of supersonic propeller airfoils and other configuration variables, the flight tests were conducted too late in the program to be of great value to an industry that was abandoning propellers and turning to turbojet powerplants. (Photo courtesy of NASA via Larry Loftin)

CHAPTER 4

THE SPACE AGENCY

When the National Advisory Committee for Aeronautics (NACA) officially became the National Aeronautics and Space Administration (NASA) on July 29, 1958, the agency made manned space exploration a primary focus of its research efforts. The numerous research facilities of NASA were immediately put to work testing new space vehicles that would transport America's astronauts on their important missions of exploration. After deciding that space capsules were the best means of providing a vehicle for manned space missions, NASA planners were faced with the dilemma of how the capsules would safely return to Earth. Extreme heat-absorbing ablative materials were developed for the capsules' re-entry into the Earth's atmosphere, but engineers examined many options for how the capsules could be steered to a safe landing on Earth. Innovative engineers at NASA Langley Research Center adapted an existing concept known as the flexible wing or "parawing" for potential applications in the space program, with extensive wind-tunnel tests and free-flying models.

NASA also pursued the development of reusable space vehicles known as lifting bodies that provided a larger operational envelope than ballistic paths with capsules. Throughout the 1960s and early 1970s, NASA conducted wind-tunnel, hydrodynamic, and flight studies of these research configurations that ultimately led to the development of the Space Shuttle. After Space Shuttle missions had become routine, NASA made plans to develop a smaller version of the Space Shuttle, which became

known as the Horizontal Lander HL-20 Space Taxi during the early 1990s. This Space Taxi was intended to transport crews to and from the International Space Station (ISS). While none of the concepts discussed in this chapter introduction ever made it into production, the research surrounding their development is fascinating and illustrates the broad scope of NASA's aeronautical research capabilities after it became "The Space Agency."

Parawing Concepts

NACA/NASA Langley engineer Francis M. Rogallo conceived the parawing concept in 1947. The parawing was an all-flexible, diamond-shaped fabric wing attached to rigid members that formed the leading edge and keel of the vehicle. Some parawing concepts were foldable and could be deployed to a semi-rigid shape for flight. Rogallo obtained a U.S. patent for his idea in March 1951. He could find no takers for the concept, with the exception of some toy stores that marketed a small version that became known as the "Rogallo Flexikite." Management within the NACA was not initially interested in the concept, but Rogallo pursued his dream of potential aerospace applications of the concept. The remarkable similarity of management's lack of interest in Rogallo's parawing and Zimmerman's flying flapjack configuration discussed earlier is worthy of note. Ultimately, the Rogallo flexible wing would become a well-known current-day sight wherever today's sport flyers conduct unpowered or powered flights.

In the early days of the space program, NASA explored many concepts for the landing and recovery of capsules returning astronauts at the end of space missions. The favored approach following reentry into the atmosphere was using parachutes for deceleration to an uncontrolled water landing and recovery. However, many leaders within the agency were interested in other, lift-producing concepts that might extend the operational mission footprint to permit landing options at land sites or runways and eliminate the complexity and cost of water landings. The Rogallo flexible-wing concept appeared as a candidate recovery system that might be deployed by the capsule crew at lower altitudes and air speeds following reentry, providing the crew a capability for gliding flight to a controlled landing.

By the early 1960s, Rogallo had finally received support from NASA to study applications of his parawing as an emergency-landing mechanism for a broad spectrum of vehicles, including Vertical Takeoff and Landing (VTOL) aircraft, re-entry vehicles, and as a drag mechanism for high-speed jet aircraft during landing. Research activities on parawing concepts began at Langley in several wind tunnels, including the full-scale tunnel; four-foot supersonic pressure tunnel; and the 300-mph 7-by-10-foot tunnel and high-speed 7-by-10-foot tunnel, which were managed by Rogallo. In addition, a series of outdoor drop-model tests of parawing/payload combinations were initiated, as well as radio-controlled flights of powered parawing configurations. A small Mylar parawing was successfully launched to an altitude of 200,000 feet and deployed in flight above the NASA Wallops Island, Virginia Flight Test Facility. Subsequently, in 1961, a large parawing successfully flew with a capsule model from an altitude of several thousand feet in a flight experiment carried out at an abandoned bombing range near Langley.

Parawing Landing Aids

One of the earliest research areas for parawing applications involved using deployable parawings to increase the lift and drag of supersonic high-speed aircraft configurations during approach and landing. In 1959 and 1960, with Rogallo's guidance, exploratory tests were conducted in the Langley 300 mph 7-by-10-foot Tunnel to determine the benefits of applying a parawing landing device to an advanced outboard-tail supersonic aircraft configuration as well as an early version of the XB-70 bomber. The parawing used for the outboard-tail evaluation had a surface area about equal to the aircraft model's wing area, and the parawing used for the XB-70 model had about twice the wing area of the basic model.

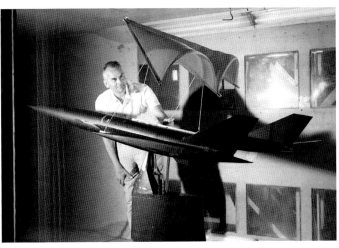

NASA Langley researcher Francis M. Rogallo inspects a parawing/outboard tail model in the Langley 300 mph 7- x 10-foot Tunnel in June 1959. Early investigations of parawings as auxiliary landing devices focused on supersonic aircraft configurations with the intent of reducing approach and landing speeds. The parawing was to be stowed within the aircraft and deployed during landing. Researchers pursued the outboard tail configuration as a particularly efficient supersonic configuration. (Photo courtesy of NASA)

Free-flight investigation of the dynamic stability and control characteristics of a concept using a parawing as a landing aid for an early version of the North American XB-70. The results of the tests in the full-scale tunnel in 1960 were positive and very encouraging. Note that in this test setup, the flexible flight cable carrying compressed air for providing thrust to the XB-70 model had to be brought onboard from beneath the model. (Photo courtesy of NASA)

Researchers anticipated that a large increase in lift could be provided by the auxiliary parawing, resulting in significant reductions in approach speed and landing requirements.

The results of the exploratory study were impressive. For example, the parawing/XB-70 configuration exhibited three times as much lift as the isolated aircraft model, and the static stability characteristics of the model were improved by adding the parawing.

Encouraged by these conventional wind-tunnel tests, researchers turned to an examination of the dynamic stability and control characteristics of such configurations using a free-flight model study in the full-scale tunnel. In the free-flight test project, another model of the XB-70 was modified with a delta-shaped parawing with an area about 60 percent greater than the model wing area. The parawing was attached to the aircraft model by several flexible riser lines. The results of the flight tests revealed that the parawing/XB-70 configuration flew much more steadily and was more controllable than the isolated XB-70 model. This result was attributed to the very large reduction in airspeed for the combination, less erratic responses to control inputs, and an increase in moments of inertia.

These exploratory studies at Langley indicated that the use of the parawing as a landing and takeoff aid appeared to be feasible from the standpoint of stability and control, and offered very large increases in lift that could be used for substantial reductions in takeoff and landing distance. However, the parawing had to be positioned high enough above the aircraft to prevent large lift losses due to undesirable mutual aerodynamic interference effects, similar to losses incurred by certain biplane wing arrangements.

Booster Recovery

Other research on the potential application of parawings in the space program was conducted at Langley in the early 1960s. A particularly innovative effort in 1963 was directed at investigating the feasibility of recovering large rocket boosters such as the first stage of the Saturn rocket with a parawing. The booster recovery concept consisted of carrying a stowed parawing on the side of the booster during the launch and boost phases of flight. At booster engine cut-off or first-stage separation, a drogue parachute would be deployed to stabilize the booster. After the booster was stabilized, the parawing would be deployed by a pilot chute at subsonic speeds, the drogue parachute would be jettisoned, and the booster would glide back under ground control to the launch site or to a down-range

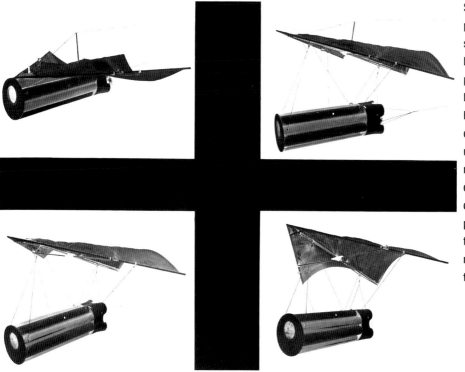

Sequence showing the final deployment of a parawing recovery system for a model of the first stage of the Saturn rocket. At upper left, the booster is stabilized by a drogue parachute and the parawing has been deployed beneath a stand-off bar that held the parawing drogue chute off the booster. At upper right, the drag of the parawing drogue has pulled the nose of the parawing upward and permitted the parawing to inflate and rise off the booster. At lower left, the parawing drogue chute has been released, and the booster drogue chute remains to stabilize the parawing/booster combination. Finally, as shown at the lower right, the booster drogue chute is released and the gliding combination is controlled to a safe landing. (Photo courtesy of NASA)

site. The Langley test consisted of flying a radio-controlled model to evaluate the deployment, dynamic stability, and control characteristics of a 0.083-scale booster/parawing configuration. The full-scale booster was about 72-feet long, 21.5 feet in diameter, and the total full-scale recovered weight (booster and recovery system) was about 120,000 pounds. The booster model carried a nose boom and instrumentation to measure airflow direction as well as cameras to record a view of the horizon and surrounding terrain.

Flight tests of the booster-recovery system were conducted at Plum Tree Island, an abandoned World War II bombing range near Langley Air Force Base, Virginia. The booster model with the stowed parawing was taken to an altitude of about 3,500 feet on a launch rig attached to the side of a helicopter and released for evaluations of the deployment analysis. For separate stability and control tests, another launch approach was used in which the model was launched in towed forward flight by the helicopter at an airspeed of about 20 knots and an altitude of 2,000 feet, with the parawing already deployed at the end of a 300-foot cable. Ground-based pilots controlled the model in both test procedures.

Results of the stability and control tests showed that the model's flight behavior was generally satisfactory after release from tow, although a small-amplitude roll oscillation occurred continuously during flight. More serious undesirable characteristics occurred when the angle of attack was increased and the parawing was stalled. The parawing-booster exhibited an abrupt stall, followed by a nose-down tumbling resulting in the booster falling on top of the parawing in an uncontrollable descent.

In the deployment tests, the research team initially attempted to install the parawing in a deployed position attached to the side of the booster. These results were entirely unsatisfactory, with frequent collisions between the booster and the parawing. However, when the parawing was folded and stowed completely within the length of the booster, the deployment sequence was satisfactory. In addition to a large drogue parachute that helped stabilize the booster during vertical descent, a smaller drogue parachute was required to extract the parawing away from the side of the booster and spread it from the booster in a satisfactory manner. This project required extensive testing of various combinations of stabilizing parachutes and parawing deployment schemes,

but a satisfactory deployment technique was developed to permit reliable gliding flight recovery for the booster.

Although the parawing/aircraft and parawing/booster application studies were interesting and stimulated numerous researchers at Langley, the center of focus and priorities for applications of parawings within the NASA program in the early 1960s centered on spacecraft recovery systems.

During the Gemini program in 1961, interest began to heighten regarding the use of parawings for land recovery of returning space capsules. As an indication of NASA's interest, industry was requested to propose candidate configurations for a parawing spacecraft-recovery system. North American Aviation was subsequently selected to develop the Gemini Paraglider concept, which used an all-flexible inflatable parawing. The project called for demonstrations of unmanned and manned gliding flights of Gemini/parawing configurations following release from tow behind a helicopter. The program encountered numerous problems, including deployment and controllability issues to be discussed later in this book, but it continued from 1961 to 1964.

The Ryan Wings

In 1961, the Ryan Co. received a contract from the Army to develop an exploratory parawing concept

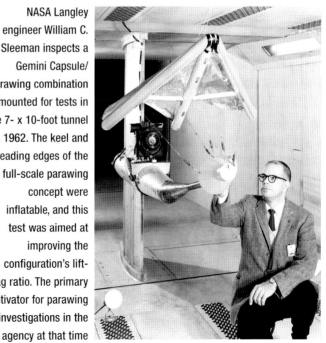

NASA Langley engineer William C. Sleeman inspects a Gemini Capsule/ Parawing combination mounted for tests in the 7- x 10-foot tunnel in 1962. The keel and leading edges of the full-scale parawing concept were inflatable, and this test was aimed at improving the configuration's lift-drag ratio. The primary motivator for parawing investigations in the agency at that time was for recovery and landing of space capsules. (Photo courtesy of NASA)

known as the Ryan Flex-Wing. Initially powered by a single pusher propeller and a 100-hp engine (later upgraded to a 180-hp engine), the Flex-Wing consisted of a simple platform/cockpit/parawing arrangement. The parawing was attached to the cargo platform by an overhead truss structure. Pilots controlled the parawing by banking or pitching the parawing with respect to the platform, and a conventional wheel and control column were used. A rudder immersed in the propeller slip-stream provided directional control. On May 23, 1961, Ryan test pilot Lou Everett made the first manned flight of a Rogallo-type parawing vehicle when he flew the Flex-Wing at San Diego. The Flex-Wing had an airspeed range of 24 mph to about 60 mph. Meanwhile, researchers at Langley had started investigations of a research configuration very similar to the Flex-Wing in order to study this type of vehicle's characteristics. Tests were underway in the 300-mph 7-by-10-foot tunnel as well as the full-scale tunnel, and an outdoor radio-controlled model of the Flex-Wing was also tested. In the full-scale tunnel, a powered 0.29-scale free-flight model of the Flex-Wing was tested in 1961. Results of the free-flight tests highlighted a deficiency in roll control at high angles of attack of the wing (above about 25

degrees) caused by adverse yaw that was inherently created by banking the wing for roll control. The rudder, however, was found to be an effective means of roll control at the high angles of attack.

Following some handling-quality issues regarding lateral control that occurred in the full-scale flight program in 1961, the Army requested that NASA assess the performance, stability, and control characteristics of the Ryan Flex-Wing aircraft in the Langley Full Scale Tunnel. Accordingly, power-off and power-on tests of the vehicle for speeds from 25 mph to 47 mph were conducted in January 1962.

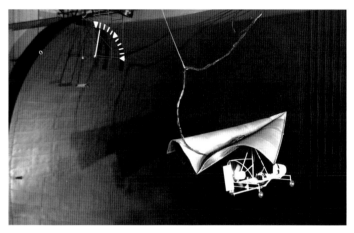

Free-flight model of the Ryan Flex-Wing flying during a dynamic stability and control test in the Langley Full Scale Tunnel in 1961. Roll control was found to be deficient at high angles of attack of the parawing due to large adverse yaw created by banking the wing for roll control. (Photo courtesy of NASA)

Radio-controlled model of the Ryan Flex-Wing vehicle being prepared for flight in 1961. Several famous Langley employees are shown, including Francis M. Rogallo (second from left with sunglasses), head of the 7-by-10-Foot Tunnels Branch; Donald E. Hewes, project leader who later became manager of the Langley Lunar Landing Research Facility (kneeling); John P. Stack, assistant director of Langley (second from right with striped jacket); and Mark R. Nichols, chief of the Full-Scale Research Division. Attendance of such high-level managers at the test site is an indication of the intense interest in parawing applications at Langley at the time. (Photo courtesy of NASA)

Ryan test pilot Lou Everett flies the Ryan Flex-Wing. Everett was also the test pilot for the Ryan Fleep parawing vehicle and the Ryan XV-5A V/STOL aircraft. (Photo courtesy of David Everett)

The results of the Flex-Wing tests showed that for parawing keel angles of less than 20 degrees, the rear of the flexible parawing experienced flutter, and for keel angles greater than 35 degrees, the vehicle exhibited longitudinal instability (pitch up). The configuration had adequate longitudinal and directional stability with

Lou Everett during high-altitude flights of the Flex-Wing. (Photo courtesy of David Everett)

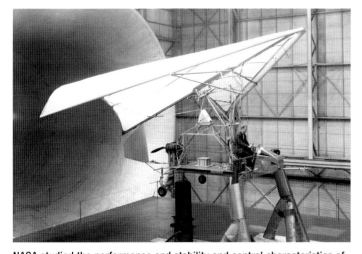

NASA studied the performance and stability and control characteristics of the Ryan Flex-Wing in the Langley Full Scale Tunnel at the request of the U.S. Army in 1962. Langley researcher John W. "Jack" Paulson poses in the pilot's seat. The full-scale tunnel tests of the Flex-Wing indicated that the configuration exhibited the same lack of satisfactory roll control at high angles of attack that was characteristic of the free-flight model. (Photo courtesy of NASA)

the rudder on, but the lateral dihedral effect was excessive, creating an undesirable effect on lateral control similar to that noted in the preceding small-scale model tests. With its original wing-banking lateral-control concept, the Flex-Wing created only small rolling moments for roll control at high angles of attack, accompanied by large adverse yaw and large lateral stick forces. This undesirable aerodynamic combination results in a reduced rolling motion when control inputs are applied and, if severe, can result in the vehicle rolling in the direction opposite that intended by the pilot. Researchers also found that the rudder was more effective as a roll-control device than banking the wing, although a time lag in the vehicle response was involved. Langley also tested a modified roll-control scheme in which the leading edge of the outer wing was hinged to permit deflections for roll control instead of banking the wing. The wingtip lateral-control arrangement had been developed at Langley during the previous small-scale model tests. This control concept produced large rolling moments and small stick forces.

The Flex-Wing experience at Ryan and Langley served as valuable guidance when the Army subsequently contracted with Ryan to develop a second-generation parawing vehicle known as the Ryan XV-8A Flexible Wing Aerial Utility Vehicle or "Fleep" (Flex-Wing Jeep). The "XV" designation was used in recognition of the STOL capabilities of the design. The Fleep vaguely resembled its Flex-Wing heritage, consisting of a cargo platform attached to a parawing by means of an overhead truss arrangement. It was also powered by a pusher propeller located at the aft end of the platform and had a cockpit at the front.

The Fleep, however, included several changes to correct deficiencies exhibited by the previous design. The wing-banking and pitching concept had resulted in unsatisfactory roll control of the Flex-Wing, and was therefore a prime concern during the Fleep's design. A new concept was designed for roll control in which wing bank was still used, but the aft 14 percent of each leading edge of the parawing was hinged to relieve the large lateral stick forces inherent in the wing-bank control system. The aerodynamic action of the deflected wingtips was similar to aerodynamic tabs used on conventional aircraft surfaces. When the wingtips were deflected differentially, they provided a rolling moment, which, in turn, banked the wing.

The Ryan lead for the Fleep design was Pete Girard, who had been the test pilot for the company's X-13 VTOL aircraft (discussed in the next chapter). Some of the Fleep design features by Ryan did not please the Langley experts, and their concerns were expressed at early meetings with Girard. Initially Ryan had proposed that the Fleep use a free-floating canard similar to that used by the Curtiss XP-55 of World War II, but Langley researchers strongly recommended against such a design, and it was dropped. A V-tail configuration was mounted at the rear behind the propeller, and control surfaces on the tail were used for pitch and yaw control. Langley was concerned that the V-tail control surfaces would be ineffective since they were removed from the propeller wake.

In 1963, a 0.33-scale free-flight model of the Fleep underwent flight evaluations in the full-scale tunnel. The longitudinal stability and control characteristics of the model were judged to be satisfactory, and roll control at low angles of attack was satisfactory. As angle of attack was increased, however, a progressive deterioration in roll-control effectiveness resulted in an unsatisfactory control condition. Disturbances caused by turbulence in the wind-tunnel test stream resulted in the model going out of control and diverging out of the test section. Once again, excessive levels of adverse yawing moments and dihedral effect resulted in a dramatic reduction in roll control at high angles of attack.

Surprisingly, the net effectiveness of roll control provided by a combination of wing bank and wingtip deflections was lower than that of the wing-bank-alone control. However, researchers concluded that the reduction in lateral stick forces with the wingtip/wing-bank controls would probably be rated higher by a pilot of a full-scale vehicle. Interestingly, the use of the V-tail control surfaces as rudders in combination with wing bank for roll control resulted in a pilot-induced lateral oscillation that quickly built up in amplitude to the point that the model diverged from the airflow in the test section. Further analysis showed that this result was caused by additional adverse yaw produced by the V-tail surfaces. When the V-tail was replaced with a conventional vertical tail and rudder in the propeller wake, as the Langley researchers had recommended prior to the project, satisfactory lateral control characteristics existed over the range of angle of attack tested.

Flight tests of the full-scale XV-8A concluded that the handling characteristics of the aircraft were good and that control harmony between the longitudinal and lateral control systems was excellent, enabling the aircraft to be flown with one hand. Stability in all cases was positive, with only light forces required. Take-off and landing performance proved the STOL capability of the airplane. At maximum gross weight, the take-off distance over a 50-foot obstacle was 1,000 feet, and landing distance to clear a 50-foot obstacle was 400 feet. Some

Free-flight investigation of the Ryan Fleep model in the full-scale tunnel. Note the V-tail configuration and centerline rudder added by Langley in the tests. Also visible are the Langley-conceived deflectable wingtips for reduced stick forces during roll-control inputs. (Photo courtesy of NASA)

Ryan's Lou Everett poses in the cockpit of the Ryan Fleep vehicle. Note the different cockpit enclosure for the second-generation parawing vehicle. (Photo courtesy of David Everett)

Front view of the Fleep during the flight-test program. Designated the XV-8A in recognition of its short takeoff and landing capability, the Fleep exhibited good stability and control. (Photo courtesy of David Everett)

test operations were conducted from unprepared desert surfaces, establishing the capability for operation from areas other than regular airfields.

NASA Langley conducted additional research using the Fleep model in 1965 to evaluate the benefits of a high aspect-ratio parawing. The relatively low aspect-ratio conical parawings discussed to this point exhibited relatively high drag, but they had aerodynamic performance that was adequate for applications in which only a moderately wide range was required. However, parawing tow vehicles or cargo-delivery configurations required a higher lift/drag ratio, which might be provided by increasing the nominal aspect ratio from three to about six. In this test program, the model's pitch and roll control were obtained by pitching and banking the wing, and yaw control was provided by a rudder surface directly in the propeller slipstream. The flying characteristics of this high aspect-ratio configuration were generally more satisfactory than those observed during the Fleep test with the aspect-ratio-three conical parawing previously discussed.

The NASA Parasev

At the NASA Flight Research Center (FRC) at Edwards Air Force Base, California, the space agency's award of a contract to North American for a Gemini paraglider recovery system in 1961 initiated a concern over the lack of handling-quality information for parawing configurations. A team of FRC researchers advocated and obtained approval for a single-seat vehicle for exploratory studies of flying characteristics. The unpowered vehicle, known as the Paraglider Research Vehicle (Parasev) was designed and constructed at the FRC, consisting of a parawing, a rudimentary seat, and a tubular structure. The pilot directly controlled the tilting of the parawing with an overhead control stick attached to the wing. Flight testing of the Parasev began in 1962. On its maiden flight, the craft was towed to altitude by a station wagon, and on subsequent test flights the craft was towed by a PT-17 Stearman biplane.

The NASA pilots disliked the lack of sensitivity in the Parasev's control system and the unconventional lags inherent to the control concept. It was judged to be a difficult vehicle to fly, and pilot Bruce Peterson was involved in a pilot-induced lateral oscillation that diverged at low speeds, resulting in a non-fatal crash. After the accident, the vehicle was rebuilt. A new control system was designed to replace the overhead control stick, using a conventional stick and pulleys to pitch and roll the parawing. The new vehicle was designated the Parasev 1-A. The Parasev continued in various flight-research programs from 1962 through 1964, completing about 350 test flights. Several famous

In 1963, flight tests of the manned Parasev parawing were initiated at the NASA Flight Research Center (now NASA Dryden Flight Research Center) at Edwards AFB, California. In this 1964 photo, the vehicle is being towed to altitude for a test flight. The purpose of this flight research program was to gather information on the handling qualities of parawing vehicles. (Photo courtesy of NASA)

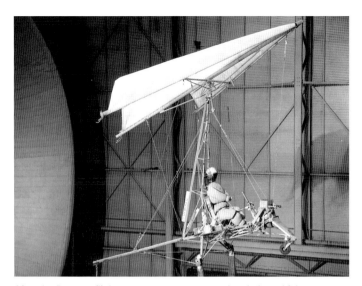

After the Parasev flight-test program was completed, the vehicle was donated to the Smithsonian's National Air and Space Museum in Washington, DC. Before it was delivered to the museum, however, it was used in a final round of generic parawing research in the Langley Full Scale Tunnel from July to September, 1964. (Photo courtesy of NASA)

research pilots and astronauts, including FRC pilot Milt Thompson, Langley pilot Bob Champine, and astronauts Neil Armstrong and Gus Grissom, flew and evaluated the Parasev. In one program it was outfitted with a smaller parawing (reduced in area from 150 square feet to 100 square feet) to evaluate the characteristics of lower vehicle lift-to-drag ratios and was redesignated as the Parasev 1-B. Finally, as the Parasev 1-C, it was outfitted with an inflatable parawing in an attempt to replicate the expected behavior of the Gemini Paraglider configuration. Unfortunately, the Parasev 1-C configuration was very unstable.

The Gemini Paraglider

As NASA conducted its fundamental parawing studies, North American was developing the Gemini Paraglider concept in an intense program. A series of mechanical mishaps had plagued the project during planned demonstration flights of 0.5-scale and full-scale unmanned vehicles. Plans had called for carrying the test capsules to an altitude of about 13,000 feet in a helicopter, releasing the test capsule, deploying the paraglider, and controlling it via radio control to a flare and landing. However, system failures frequently occurred, preventing deployment, and concern about the concept's viability quickly rose. NASA's plans for

incorporating the paraglider/capsule concept in the Gemini program was delayed to Gemini 9; then on June 12, 1964, NASA announced its decision to eliminate the Gemini paraglider concept from the Gemini program. NASA did, however, encourage North American to use the equipment for further paraglider research.

On July 29, 1964, the first manned captive-flight test of the Gemini Paraglider took place. A helicopter towed the test vehicle to an altitude of about 2,000 feet. Following a 20-minute towed flight, the tow was completed down to a landing and the towline released successfully. On August 7, a manned free-flight test at Edwards Air Force Base ended in a dramatic failure. In this test, the vehicle was towed by a CH-47 helicopter to an altitude of 15,500 feet and released. After it was released, the vehicle went into a series of uncontrollable yawing oscillations and drifted toward the active Edwards runway. Without response from his control inputs, test pilot E. P. Hetzel bailed out using an emergency parachute.

Researchers at Langley were brought in as consultants on the manned accident, and when they tested the parawing/capsule configuration in glide tests at Langley, they found that the uncontrollable yawing oscillations could be duplicated. The motions appeared to be a phase problem with the capsule yawing to the right, followed later by the parawing yawing to the right. By that time, the capsule had yawed back to the left, and an unstable oscillation ensued. The Langley researchers re-rigged the configuration with additional lines between the capsule and parawing to prevent the out-of-phase motions, and the modified design conducted very successful gliding flights at the Plum Tree Island range near Langley. In those flights, the configuration flew very well and displayed adequate control. North American subsequently conducted additional 0.5-scale and full-scale unmanned radio-controlled flights with another parawing design, and a piloted full-scale flight from release was successfully completed on December 19, 1964. North American continued to conduct exploratory flights of the concept into 1965, including a successful manned flight from tow release to landing by a test pilot named Jack Sweigert, who would later become a NASA astronaut and achieve fame in the Apollo 13 mission.

In any event, the 1964 NASA decision to terminate the paraglider space-vehicle-recovery-system program in favor of conventional parachute systems immediately decreased the level of research on parawing applications in the agency.

Despite the lack of NASA applications, Francis Rogallo's parawing concept lives on today as a worldwide favorite of sport aviation in the form of powered and unpowered hang gliders. Spurred on at the end of the 1960s by initial applications within the water-skiing community, a new form of fun sport had been born, and it remains strong to this day with land- and water-based applications.

Lifting Bodies

Early in the NASA space program, technical candidates for reentry and landings of spacecraft were identified to extend operational range and landing options. Capsules and other symmetric spacecraft shapes designed for the U.S. moon mission would follow a ballistic trajectory during reentry, which minimized potential downrange flexibility and options to seek landing sites other than water. In contrast, vehicles capable of providing lift for maneuvers during reentry would provide crews the ability to deviate from a ballistic path and glide to a safe land-based runway landing like an aircraft. The prominent NASA leaders in the conception and development of "lifting bodies" in the 1950s were H. Julian Allen and Alfred Eggers of the NASA Ames Research Center. Led by Allen and Eggers, engineers at Ames began conceptual studies of blunted half-cone shapes that could be used as potential lifting bodies. Although the lift-to-drag ratios produced by the configurations (about 1.5 at hypersonic speeds) were much lower than those associated with aircraft, the lift would provide a revolutionary capability for control of reentry flight parameters. The subsequent evolution of the lifting-body concepts resulted in exciting new radical configurations requiring major research activities at the Langley, Ames, and Dryden research centers.

Evolution of the M2

The early Ames conceptual studies led to the development of a family of potential reentry configurations. One of the first designs, known as the Eggers M1, resembled a sawed-off rocket nose cone. Researchers in the full-scale tunnel at Langley were following the development of the Ames configurations with an interest in determining the dynamic stability and

control characteristics of such radical configurations. In 1959, free-flight model tests of the M1 design were conducted with very positive results, further increasing the growing interest in lifting bodies. The Ames staff continued to refine and further develop their configurations, resulting in a half-cone shape with a blunted 13-degree nose semi-apex angle known as the M2. As the concept continued toward maturity with additional wind-tunnel testing over the operational speed range, major deficiencies in stability and control were noted at subsonic speeds, requiring modifications to the aft end of the configuration. The M2 lifting body configuration included stubby vertical fins and horizontal elevons, with trailing-edge flaps on the body to provide longitudinal trim.

Meanwhile, at the NASA Flight Research Center at Edwards, researcher Robert D. "Dale" Reed was following the rapidly evolving M2 concept with great interest. A preeminent innovator, Reed used a series of radio-controlled model tests to convince management of the M2 configuration's feasibility and controllability at low subsonic speeds. Following this success, Reed and his associates gained approval for the construction and flight testing of a piloted unpowered glider of the M2 to assess the flying characteristics of the lifting body at subsonic speeds and during landings. The research-vehicle concept was designed to be extremely low cost, built of

Free-flight model of the Ames research center's M1 conceptual lifting-body configuration is prepared for flight tests in the Langley Full Scale Tunnel in 1959. The results of the flight tests were very positive and increased emerging interest in the potential capabilities of lifting bodies to provide extended operational footprints for landing options of returning spacecraft. (Photo courtesy of NASA)

a plywood shell over a tubular steel frame and with a simple control system. It was envisioned that the unpowered vehicle, known as the M2-F1, would be towed glider-fashion to altitude and released for the glide-test evaluations. In typical fashion, the ingenious M2-F1 team procured and modified a Pontiac Catalina for the initial tow tests.

At Langley, wind-tunnel support for the development of the M2-F1 was provided in 1960 by conventional stability and control investigations of a 0.34-scale model in the Langley 12-foot Low-Speed Tunnel (previously the NACA Free-Flight Tunnel) and the Langley Full Scale Tunnel. In addition to tests to define quantitative aerodynamic data for lift, drag, and stability and control, free-flight tests were made in the full scale tunnel to determine the dynamic stability characteristics of the early M2-F1. To adapt the model of the unpowered M2-F1 to the level-flight testing technique used in the full-scale tunnel, a compressed-air jet at the rear of the model was used for thrust. The results of the free-flight tests showed that the model had satisfactory longitudinal characteristics, but a lightly damped lateral oscillation was observed at high angles of attack. The oscillation was, however, completely damped out with the use of a roll-rate feedback system.

In 1962, Langley conducted another series of free-flight model tests in the full-scale tunnel in support of the M2-F1. In this tunnel entry an updated version of the 0.34-scale model was tested in towed flight to determine its dynamic stability and control characteristics prior to the full-scale flight tests at the NASA Flight Research Center. For the tow tests, the model was equipped with artificial roll-and yaw-rate dampers. The test set-up used an aircraft cable attached to the wind-tunnel turning vanes ahead of the tunnel contraction. An overhead cable similar to the normal set-up used in free-flight model tests supplied electric power and control signals for the flights. The model was tested both with and without a center fin and a canopy.

Results of the tow tests indicated that the longitudinal stability characteristics of the M2-F1 model were

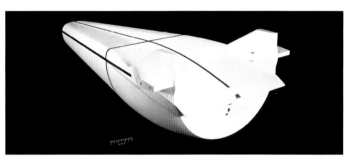

Rear view of the Langley free-flight model of an early M2 configuration. All-movable surfaces on the outer sides of the stubby vertical tails provided roll and pitch control, while the flap on the aft body was used for trim. (Photo courtesy of NASA)

The M1 free-flight model undergoing flight evaluations in the full-scale tunnel. Note the flexible flight cable used to bring compressed air onboard the model for thrust to permit level flight and the tab-like control surfaces. By 1959, the Langley free-flight technique had been moved from the free-flight tunnel to the full-scale tunnel, and the flight crew was seated outside the tunnel airstream in an unenclosed balcony. (Photo courtesy of NASA)

Free-flight testing of the M2 model required a modified approach to the traditional free-flight technique with the flexible flight cable entering from the underside of the model. Results of the flight tests showed that the configuration exhibited a lightly damped roll oscillation at high angles of attack, but a conventional roll damper heavily damped the motions. (Photo courtesy of NASA)

satisfactory. The lateral stability and control characteristics, however, were found to be deficient without artificial damping. The model exhibited a lightly damped roll oscillation that was easily excited by pilot control inputs; however, the pilot could control the motion with coordinated rudder and aileron control. The model also displayed an unstable long-period lateral oscillation referred to as a towline oscillation. The period of the towline oscillation was much longer than those of the roll oscillations, permitting the pilot adequate time to control the model. If the motions were not controlled, the model diverged out of the tunnel test section, and the test was terminated. The results also indicated that large unfavorable yawing moments accompanied the use of the outboard horizontal tails as ailerons for roll control. Elevons alone were found to provide acceptable roll control. In its basic configuration, the model in towed flight exhibited characteristics that the Langley staff regarded as unsatisfactory except for relatively short periods of towed flight. The addition of artificial damping in roll and damping in yaw resulted in a very stable towed configuration.

The first flight tests of the M2-F1 occurred at the Flight Research Center on April 5, 1963. The research vehicle was towed behind the souped-up Pontiac and showed generally good characteristics during the flights; however, test pilot Milt Thompson encountered lateral oscillations during the flight attributed to an inadvertent control-mixing problem. Later flights of the M2-F1 used a C-47 as a tow plane for in-flight tows to higher release altitudes.

The successful flight testing of the M2-F1, essentially a light-weight, low-cost concept demonstrator, gave NASA the confidence to proceed with a pair of more sophisticated heavyweight rocket-powered lifting bodies for evaluations of handling qualities of more representative lifting-body configurations across the speed range from supersonic flight to landing. The craft would be launched from a NASA B-52 mother ship and be powered by XLR-11 rocket engines. These two configurations were known as the M2-F2 and the HL-10. Following a NASA review of competitive industry designs, Northrop built both vehicles. After construction and initial check-out, the vehicles were first tested in the huge 40- by 80-foot Wind Tunnel at the Ames Research Center prior to flight.

Free-flight model of the M2-F1 installed in the full-scale tunnel in 1962 for tests to determine dynamic stability and control characteristics during towed flight. Author Joseph R. Chambers, who began his career at the Langley Full Scale Tunnel during the week the tests were conducted, observes the test setup. Note the towline attached to the wind-tunnel turning vane structure in the entrance cone, and the flight cable consisting of a steel aircraft cable for snubbing the model and an electrical control-signal cable. (Photo courtesy of NASA)

The M2-F1 lifting body in towed flight behind a NASA C-47 at the NASA Flight Research Center in October 1963. The M2-F1 conducted more than 400 tows behind a souped-up automobile and 77 tows behind aircraft. (Photo courtesy of NASA)

Evolution of the HL-10

In 1962, researchers at NASA Langley, led by Eugene S. Love, conceived another radical lifting body known as the Horizontal Lander (HL) –10. In contrast to the half-cone designs of the M2 series, the HL-10 was a flat-bottomed, inverted-airfoil shape with a split trailing-edge elevon for pitch and roll control. Tip fins and a center fin were added to increase directional stability, and a series of vertical tail configurations were evaluated. The initial design objectives for the HL-10 included goals for a trimmed hypersonic lift-to-drag ratio of 1 without elevon deflection (to avoid local heating problems), and a subsonic lift-to-drag ratio of 4.0 for approach and landings.

The extent of research conducted at Langley on the HL-10 was massive. In addition to analyses of reentry trajectories and heating environments, design studies of internal and personnel layout, landing-gear design, and geometric shaping were conducted. Virtually every wind tunnel at Langley tested the configuration. Models of the HL-10 underwent aerodynamic, heating, launch-vehicle compatibility, dynamic stability, and ground- and water-landing tests; and piloted simulator evaluations were also conducted based on aerodynamic inputs from the tunnel testing.

In 1964, the low-speed aerodynamic characteristics of the HL-10 were investigated in the full-scale tunnel with a 60-inch-long free-flight model powered with compressed air. Langley researchers found that the design possessed excellent stability and control characteristics, particularly at the angles-of-attack needed for approach, flare, and landing. In fact, at the low speeds of the tests, the model was controllable to angles-of-attack as high as 45 degrees, well in excess of the maximum value of 25 degrees predicted for approach, flare, and landing. Rolling oscillations that had been noted in free-flight tests of other highly swept shapes were well damped for the three-finned HL-10 configuration.

The HL-10's landing characteristics, including ground-runout behavior, were extensively studied by researchers, using the impact structures facility at Langley. These studies included tests of HL-10 models equipped with twin main skids and a nose wheel for conventional ground landings. For the conventional ground-landing tests, a scaled HL-10 model was catapulted across the ground surface. Researchers found that

the vehicle possessed a tendency to tumble at touchdown. The tumbling effect was caused by an excessive roll attitude, which could be alleviated by incorporating a steerable limited-torque nose wheel in the design. Emergency landings using parachutes were also studied, as were water landings. For emergency water landings, the best results were obtained when a parachute was deployed from the vehicle for a near-vertical descent in which the design hit the water tail first. In addition to the landing tests, heat-transfer and pressure-distribution studies on the HL-10 were conducted in the Langley "Hotshot" tunnel, with tests performed at Mach numbers reaching 20. The tests showed favorable results for the design.

One major aerodynamic area of concern during the HL-10 tunnel testing was the discovery that a severe degradation in directional stability occurred at low supersonic speeds. Piloted simulator studies showed that the vehicle would have unsatisfactory handling qualities and that configuration modifications were required to increase directional stability. Following several subsequent wind-tunnel tests that investigated the effects of variations in the geometry of the tip fins, researchers arrived at a configuration that increased the fin area, toe-in angle, and roll-out angle. Throughout the study, researchers were sensitive to the requirement to increase stability without reducing aerodynamic performance. After all this research was accomplished, the HL-10

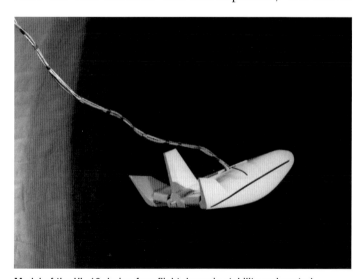

Model of the HL-10 during free-flight dynamic stability and control evaluations in the Langley Full Scale Tunnel in 1964. Note the inverted-airfoil shape of the configuration, the blunt trailing edges of the elevon surfaces, and the auxiliary thruster tube powered by compressed air for level flight. (Photo courtesy of NASA)

vehicle finally evolved as a triple-fin configuration, based on the massive amount of data that had been generated with no less than 10 different HL-10 models in various wind tunnels. More than 8,000 hours of tunnel testing was involved, and the model tested in the Langley Full Scale Tunnel was even larger (28-feet long) than the actual vehicle (20-feet long).

Langley had previously conducted critical wind-tunnel testing of the launch-separation dynamics for the X-15 research aircraft and the B-52 mother ship in the Langley High-Speed 7-by-10-foot tunnel. Those tests had proven to be invaluable contributions to risk reduction during the flights of the X-15 at the Flight Research Center. Once again, the possibility of a collision between the B-52 and either the HL-10 or M2-F2 required analysis. In 1965, researchers used 0.025-scale models of the M2-F2 and B-52 for measurements of aerodynamic interference factors that would influence the path of the lifting bodies following launch. In turn, engineers at the Flight Research Center used the aerodynamic data in a simulator to evaluate the trajectory of the launch path. Analysis of the results by the NASA team indicated that a potential collision problem existed for the M2-F2 when launched from the original B-52 pylon used for the X-15 drops. The B-52 pylon was subsequently redesigned to lower the M2-F2 and

provide more clearance between the fins of the M2-F2 and the attachment pylon. Successful drops of both the M2-F2 and HL-10 followed.

Model of the HL-10 mounted in the NASA Langley Full Scale Tunnel for low-speed aerodynamic testing. This 28-ft model was actually larger than the full-scale vehicle. Configurations studied in the test included the body alone, the body plus center fan, and the body plus center and tip fins. Stability and control characteristics were found to be satisfactory, and tuning of the fins, rudder, and elevons dramatically improved the subsonic lift-drag ratio. (Photo courtesy of NASA)

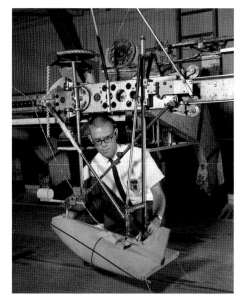

Landing behavior of the HL-10 during ground and water landings was investigated at Langley. In this 1963 photo researcher Sandy Stubbs inspects a model prior to a ground-landing study. (Photo courtesy of NASA)

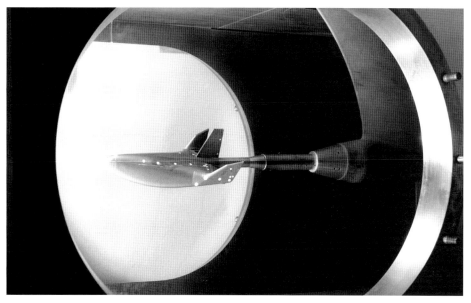

The HL-10 vehicle was conceived and developed in the Langley Aero-Physics Division (APD) using theoretical and experimental techniques. Tests in APD facilities encompassed Mach numbers from 6.8 to 20 and included pressures, heat-transfer, stability, control, and performance investigations. This photograph shows the test setup in the APD Mach-10 Continuous Flow Hypersonic Tunnel. (Photo courtesy of NASA)

In a truly remarkable accomplishment, the skilled Northrop team built the M2-F2 and the HL-10 for $1.2 million each, when industry observers had estimated that the vehicles would cost more than 10 times that amount. Because the tail geometry of the HL-10 required modifications during the construction activities to resolve stability problems, the M2-F2 was completed first.

Following wind-tunnel tests in the Ames 40-by-80-Foot Tunnel, the M2-F2 began its flight test program at the Flight Research Center with the first glide flight on July 12, 1966. By the end of the year, four pilots had conducted 14 flights in the M2-F2.

After its delivery to the Flight Research Center by Northrop, the HL-10 underwent full-scale tunnel testing at Ames. The HL-10's maiden unpowered flight occurred on December 22, 1966, with NASA pilot Bruce Peterson at the controls. The first flight, which was launched from the B-52 at an altitude of 45,000 feet, proved to be very exciting, and it was a test of Peterson's piloting skills. Both pitch and roll sensitivities were encountered, and the vehicle exhibited limited-cycle oscillations that complicated the timing of pilot inputs, resulting in marginal controllability in pitch and roll. Because of lack of response to control inputs at high angles of attack, Peterson maintained a low angle of attack to maximize control effectiveness, resulting in a very high-speed approach and landing touchdown at more than 310 mph, one of the fastest landing speeds of any piloted vehicle in history. These deficiencies were obviously unacceptable to the flight-test team, and Langley was asked for support to provide solutions to the problems. Subsequently, the HL-10 program was put on hold for 15 months while more analyses and wind-tunnel tests were conducted.

At Langley, a team from the Flight Research Center joined with the wind-tunnel staff of the 7-by-10-foot High Speed Tunnel to plan a recovery effort for the project. The team concluded that the flight problems had been a direct result of flow separation at the juncture of the tip fins and the fuselage, which in turn had negative effects on the control surfaces of the HL-10 vehicle. The elevon effectiveness had actually been higher in flight than that expected from the preliminary full-scale wind-tunnel tests at Ames, resulting in over-sensitivity of control inputs. Interestingly, a small

The Langley-conceived and -managed HL-10 on the lakebed. The HL-10 flight-test program resulted in the successful completion of 37 test flights, making it one of the most promising lifting-body designs under consideration for possible use as a space shuttle. (Photo courtesy of NASA)

After demonstrating success in pre-flight analysis of the launch dynamics of the X-15 from the B-52 carrier aircraft, the Langley staff at the 7- by 10-foot tunnel was requested to conduct a similar analysis for the M2-F2 and HL-10 vehicles. This photograph shows the M2-F2 model mounted on a variable-position test rig on a B-52 model to acquire aerodynamic data used as inputs in a launch trajectory analysis. Results of the test showed that use of the X-15 launch pylon would probably result in the fins of the M2 striking the pylon during the launch. After testing and analysis were completed, the pylon was modified to lower the M2 relative to the original position. (Photo courtesy of NASA)

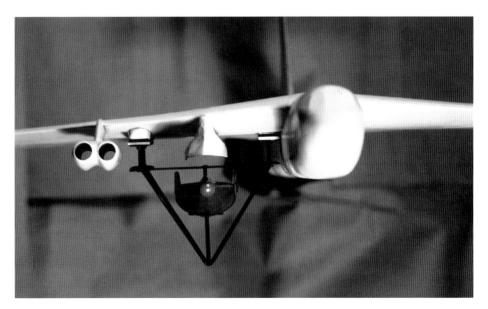

0.063-scale model at Langley had earlier provided correct predictions, but the full-scale tests had been accepted as being more accurate. Consequently, further wind-tunnel tests at Langley in 1967 were conducted in several facilities, which led to options for redesign of the tail fins. The final modification included a cambered leading edge to the fins that prevented the flow separation phenomenon.

Flight Testing

The M2-F2 program continued with flight tests into 1967. The program came to a halt on May 10, 1967, when NASA test pilot Bruce Peterson experienced severe roll oscillations before making his landing approach in the vehicle at Edwards. Experiencing roll rates as high as 200 degrees per second, Peterson became disoriented and concerned over a rescue helicopter that appeared in his flight path. In addition, Peterson failed to extend the M2-F2's landing gear in time, and the vehicle tumbled end over end the lakebed. Peterson suffered serious injuries in the accident. This incident was the only serious accident in the lifting-body test program, which included 12 and one-half years of flight tests and eight different lifting bodies. NASA engineers at Ames conducted tunnel testing, and the vehicle was modified by adding an additional center vertical fin. The renovated M2-F2 was designated the M2-F3 and went on to successfully demonstrate that "pin-point" unpowered steep approach or "dead stick" landings could be successfully made in a reusable-type lifting-body vehicle.

Flight tests of the HL-10 with the tail modifications began on March 15, 1968. Visual observations of the flow over the rear of the vehicle during the flight using wool tufts indicated that the redesign had solved the

NASA test pilot Bill Dana (right) observes a flyover of the NASA NB-52 mother ship after performing a successful test flight in the NASA Langley-conceived HL-10 on May 20, 1969. Dana flew the HL-10 to an altitude of 90,030 feet, the highest altitude reached in the lifting-body program. (Photo courtesy of NASA)

separation problem, and the handling qualities were very satisfactory. The final HL-10 configuration was regarded by all test pilots as having excellent flying qualities, and it completed its subsequent flight testing with outstanding success. HL-10 flights continued throughout the late 1960s and into 1970 with the vehicle setting speed (Mach 1.86) and altitude (90,303 feet) marks in 1970, becoming the fastest and highest-flying lifting body. Between 1966 and 1970, the HL-10 completed 37 research flights.

During the latter 1960s, the U.S. Air Force had joined the lifting body research program, and yet another lifting body, the Martin X-24A, was developed. Referred to as the "Flying Football" by some of its pilots, the X-24A added

A lineup of three famous lifting-body research vehicles. Pictured from left to right are: the Martin Marietta X-24A, the Northrop M2-F3, and the Northrop HL-10. Although none of the lifting body configurations was selected for the Space Shuttle design, the tremendous scope of data gathered in their extensive research programs proved vital to the development of the Space Shuttle. (Photo courtesy of NASA)

more proof to the fact that "pin-point" unpowered steep approach or "dead stick" landings could be successfully made with such a vehicle. Wind-tunnel tests to study the separation characteristics of X-24A and B-52 models were conducted at Langley and provided valuable data prior to flight tests of the vehicle. The last of the lifting bodies, the Martin X-24B, was a delta-shaped version of the X-24A and also demonstrated favorable high-speed lifting body qualities in flight.

Although none of the lifting-body designs were ever adopted as the eventual Space Shuttle design, they provided a wealth of knowledge and data that contributed to the development of the orbiter. This included the experience and knowledge of demonstrating how to make "pin-point" unpowered steep approaches and "dead stick" landings in a reusable re-entry type vehicle.

Martin SV-5D Prime Project

During the mid-1960s, the United States Air Force sought the development of a manned lifting body reentry vehicle capable of being launched into space and returning through a gliding reentry into the Earth's atmosphere. The vehicle, designated the Space Vehicle 5 (SV-5) Prime (Precision Recovery Including Maneuvering Entry), was built by Martin and designed to test new heat-absorbing ablative materials required for atmospheric reentry as well as the aerodynamics of such

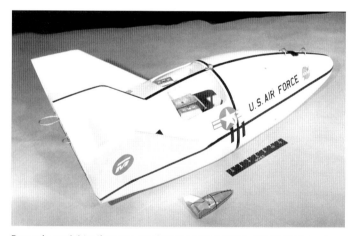

Dynamic model testing was conducted in the Langley Full Scale Tunnel with these two models of the SV-5 configuration to determine the dynamic stability of the configuration in towed flight following an aerial snatch recovery by a C-130. Langley researchers determined recommended tow attachment points on the models and demonstrated the ability to reel the towed smaller model into the cargo bay of a model C-130. (Photo courtesy of NASA)

a vehicle. The SV-5 configuration was to serve as the basis for the X-24A lifting-body design. The SV-5 was launched atop an Atlas rocket booster and retrieved by a JC-130B transport via aerial "snatch" following atmospheric reentry. The lifting body, with parachute deployed, was snatched during its descent by a special grabbing mechanism outfitted on the JC-130B. The parachute, with vehicle attached, was then retrieved aboard the JC-130B.

At the request of the Air Force in 1965, NASA conducted an investigation of the low-speed dynamic stability of the SV-5 in towed flight in the Langley Full Scale Tunnel. The investigation used two dynamically scaled flying models at 0.71- and 0.097-scale. A one-fifteenth-scale model of a C-130 was also used in the tests of the small model. The C-130 model was not, however, used in tests with the large model, and the large model was not tested with a deployed parachute.

The tests were aimed at determining the dynamic stability of the SV-5 configuration in towed flight following aerial snatch. During the tests, the effects of various towline attachment positions on the models were analyzed for various flap settings of the SV-5. The C-130's wake and its impact on the towed lifting body were also evaluated. The results of the tests indicated that there were a total of six optimum tow-attachment positions. Of these six positions, four were located in the front of the SV-5 design and two on the tail of the vehicle. Tests determined that the tail-tow attachment positions were less affected by disturbances. Flap deflections of the vehicle during towed flight resulted in degraded behavior. Langley researchers also found that the C-130's wake did not create an unacceptable problem, as the lifting body was reeled aboard and retrieved within the cargo bay.

SV-5D vehicles later completed successful flights following launches from Vandenberg AFB, California, in 1966 and 1967. During these flights, a wealth of information was obtained regarding ablative materials vital for a manned lifting reentry vehicle, in addition to the aerodynamic configuration required for a successful design. The aerial snatch-retrieval method failed during the first two missions, but success was finally achieved on a third mission conducted on April 20, 1967. The mission's conclusion was marked by the successful aerial snatch and retrieval of the vehicle at an altitude of about 10,000 feet. Air Force and NASA officials were elated, and the Prime program reached a positive end.

NASA Langley's expertise in dynamic stability simulation and testing played a vital role in helping to bring the Prime program to a successful conclusion.

The HL-20

During the early 1990s, NASA revisited the lifting-body concept when Langley proposed the development of its conceptual Horizontal Lander (HL)-20 lifting-body design. Initially known as the Personnel Launch System (PLS), the HL-20 was intended as a reusable re-entry space vehicle that could serve as a "Space Taxi," ferrying crews to and from the International Space Station (ISS). One of the design constraints was that the vehicle was required to fit entirely in the payload bay of the Space Shuttle. It was hoped that the HL-20 could relieve the Space Shuttle from this mission and even serve as an emergency escape vehicle from the ISS should a catastrophic event occur.

The HL-20 was to be piloted by two crewmembers and could seat eight other people. The vehicle would be launched atop a Titan IV booster equipped with solid rocket motors. Upon completion of its mission, the HL-20 would re-enter the Earth's atmosphere and glide to a landing like the lifting bodies and the Shuttle. The design was to serve as a low-cost alternative to the Space Shuttle. Plans were made to equip the vehicle with parachutes in case a mission had to be aborted.

Wind-tunnel investigations of the aerodynamic characteristics of the HL-20 design began in earnest at NASA Langley. Numerous models were constructed and tested in Langley's wind tunnels from subsonic to hypersonic speeds. The wind-tunnel tests revealed that the design was stable at all speeds. Re-entry heating investigations in wind tunnels were also performed using HL-20 models.

By the 1990s, NASA had developed advanced computational fluid dynamics (CFD) technology to supplement

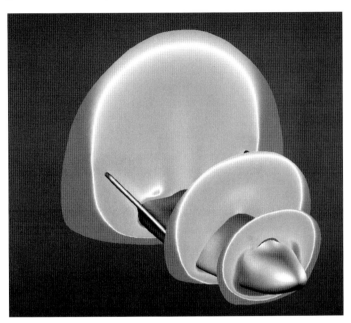

The technology of computational fluid dynamics had rapidly matured during the years after the lifting-body programs of the 1960s. This graphic shows a representative CFD result for Mach contours around the HL-20. These computational efforts were conducted in tandem with extensive wind-tunnel tests at Langley. (Photo courtesy of NASA)

Artist rendering of the HL-20 vehicle performing a mission from the International Space Station (ISS). Piloted by a two-person crew, the HL-10 was designed to carry eight passengers and would serve as a low-cost alternative to the Space Shuttle. (Photo courtesy of NASA)

NASA, North Carolina State University, and North Carolina A&T University partnered to build a full-scale version of the HL-20 for engineering analyses. Shown on October 22, 1990, the mock-up was used for human-factors studies, including ease of entry and egress. (Photo courtesy of NASA)

its wind-tunnel studies. Langley had become a leader in the use of such technology to investigate, simulate, and predict aerodynamic flows, shock-wave interactions, and aerothermodynamic heating for space vehicles, and used this computational expertise to further the development of the HL-20 design.

An HL-20 flight simulator was also developed at Langley that enabled pilots to practice flying the vehicle in the approach and landing portions of a mission. Through the use of the simulator, pilots determined that safe and accurate landings could be made in the HL-20.

Lockheed initiated an HL-20 prototype practicality study in October 1991. As part of the activities, NASA, North Carolina State University, and North Carolina A&T University partnered to build a full-scale version of the HL-20. Numerous human-factors studies were conducted to examine crew entry and egress, effects of the number of individual crewmembers, accommodations, and visibility. The study, which ended in December 1991, concluded that a 10-member crew could easily enter and exit the vehicle while in a horizontal attitude. Entry and exit from the vehicle in a vertical attitude presented more of a challenge as crews had to enter or exit through a hatch and accompanying ladder. The human-factors studies brought about minor refinements to the HL-20 design. In the years that followed, assembly of the ISS commenced, and space planners decided that a Russian Soyuz spacecraft would serve as the primary emergency escape vehicle. Consequently, development and production of the HL-20 was discontinued.

The Lunar Landing Training Vehicle

When President John F. Kennedy committed the United States to a moon landing in his 1961 speech, NASA researchers were shocked with the urgency of the project and the complex technical issues it entailed. When the decision was made to accept a lunar rendezvous technique using a lunar excursion module for landing on the moon, one of the foremost issues was how to design a vehicle for the landing task, and how to train astronauts to a satisfactory level of proficiency for the mission. Bell Aerosystems was a major participant in the development of vertical takeoff and landing (VTOL) aircraft technology in the 1950s, and company personnel were well acquainted with controllability and VTOL performance issues. Therefore, Bell was in a position to react quickly to an initial request from NASA in late 1961 for concepts for a free-flying piloted simulator for astronaut training. Hubert M. Drake, who started his NASA career as a researcher conducting free-flight model tests at the Langley Free-Flight Tunnel, had conceived the idea.

In May 1963, a 0.30-scale model of the original design for a lunar-lander training vehicle was tested in the 17-foot test section of the Langley 300-mph 7-x-10-Foot Tunnel. The test program included a jet simulator to replicate the effects of the jet engine. Not surprisingly, the measured aerodynamic data for the unconventional tubular structure differed considerably from Bell's predictions. The results also indicated that the aerodynamic forces on the vehicle were considerable.

NASA Lunar Landing Research Vehicle (LLRV) in flight with chase helicopter in April 1965. The first two LLRVs were configured with the open cockpit shown in the photograph. (Photo courtesy of NASA)

There would subsequently be cause to reflect on these results after a future training accident.

Bell subsequently designed and delivered two Lunar Landing Research Vehicles (LLRVs) to the NASA Flight Research Center (now the NASA Dryden Flight Research Center) in 1964. The spindly, flying bedstead resembled early VTOL prototypes, consisting of tubular truss construction and an open cockpit with an ejection seat. A turbofan engine mounted vertically in a gimbal provided vertical lift in excess of vehicle weight to propel the craft to a predetermined test altitude. The thrust from the engine was then reduced to support five-sixths of the vehicle's weight, representing the effects of the moon's gravitational field. Hydrogen peroxide lift rockets were used to modulate rate of descent and translational movement during flight. Thrusters powered by hydrogen peroxide were used to provide pitch, roll, and yaw control.

The first flight of the two LLRVs occurred at Dryden in October 1964, and after two years of very productive flight research and pilot training, NASA contracted with Bell to deliver three updated Lunar Landing Training Vehicles (LLTVs). These vehicles had improved capabilities to simulate the Apollo Lunar Excursion Module (LEM) features, including a cockpit display and control system representative of the LEM. One very significant modification to the original LLRV design consisted of arranging the cockpit to simulate the enclosed cockpit of a lunar module. The modified configuration had three walls and a roof around the cockpit, with only the front area exposed.

On December 8, 1968, NASA pilot Joe Algranti (chief of aircraft operations at NASA Johnson Space Center) conducted a flight of LLTV 1 (NASA 950), which had conducted 282 flights, at Ellington Air Force Base, Texas, as an acceptance flight following vehicle modifications. During the flight, the vehicle abruptly yawed off to the side, and Algranti's control inputs could not stop it. The vehicle rapidly rolled from 90 degrees to the right to 90 degrees to the left. Algranti safely ejected from the vehicle only three-tenths of a second before impact. A quick review of the flight data revealed that more control was needed than the yaw thrusters could provide. The ensuing accident investigation board recommended that another LLTV be tested in the NASA Langley Full Scale Tunnel to investigate the configuration's aerodynamic behavior and controllability. Accordingly, LLTV 3 (NASA 952) was loaded aboard a modified Boeing Stratocruiser and flown to Langley for installation in the tunnel. The modified Stratocruiser, known as the Super Guppy, had been used to transport the upper stages of the Saturn rocket to Cape Canaveral.

The test program in the Full-Scale Tunnel in January 1969 was planned to include both unpowered and powered tests. The unpowered test results indicated that the modified cockpit configuration of the LLTV caused large unstable aerodynamic yawing moments when the vehicle was sideslipped as little as two degrees. Basically, the cockpit arrangement acted as a large "sugar scoop" to destabilize the vehicle beyond controllability limits. Researchers then modified the cockpit by removing the roof, thereby venting the destabilizing area. No power-on testing was accomplished in the tunnel tests because of excessive vehicle vibrations.

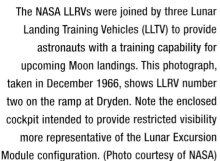

The NASA LLRVs were joined by three Lunar Landing Training Vehicles (LLTV) to provide astronauts with a training capability for upcoming Moon landings. This photograph, taken in December 1966, shows LLRV number two on the ramp at Dryden. Note the enclosed cockpit intended to provide restricted visibility more representative of the Lunar Excursion Module configuration. (Photo courtesy of NASA)

Based on the tunnel data, the flight-training program adopted a preliminary restricted flight envelope for angle-of-attack, speed, and angle-of-sideslip. After validating the flight envelope with additional flight tests, the training continued and the yaw-control problem was resolved. Soon thereafter, Neil Armstrong conducted several training sessions and pro-ficiency flying in the LLTV before the Apollo 11 mission on July 16, 1969. The LLTV (NASA 952) that was used for the tunnel tests at Langley is now on exhibit at the Johnson Space Center in Houston, Texas.

Thanks to the availability of an old wind tunnel and its dedicated staff, NASA had resolved a major problem in the Apollo Program.

Arrival of the LLTV at Langley Air Force Base in January 1969 in preparation for aerodynamic tests in the Langley Full Scale Tunnel. Unloaded from the Super Guppy transport, the LLTV awaits transfer to the tunnel. (Photo courtesy of NASA)

Another view of the LLTV at Langley showing the enclosed cockpit, which created large yawing moments in a sideslipped condition, as might be caused by sudden crosswinds. The magnitude of the moments exceeded the control power available to the pilot. (Photo courtesy of NASA)

Test setup for the full-scale tunnel investigation of the aerodynamic behavior of the LLTV in January 1969. Results of the tests identified uncontrollable aerodynamic moments created by the enclosed cockpit as the probable cause of the LLTV accident that occurred at Houston the previous month. Subsequently, the cockpit roof was removed and the unacceptable condition eliminated. (Photo courtesy of NASA)

CHAPTER 5

UP, UP, AND AWAY

The scientific interest in aircraft with a vertical takeoff and landing capability had heightened during World War II, but the relatively heavy piston engines used at the time severely limited the consideration of designs to rotorcraft, with an emphasis on helicopters. However, the introduction of turbine engines such as the turbojet and the turboprop for conventional aircraft resulted in a significant increase in the power-to-weight ratio of engines, stimulating designers to take a fresh look at vertical-takeoff configurations with the new higher performance potential.

The emergence of these new powerplant concepts and the demonstrated utility of the helicopter in the Korean War spurred on a multitude of studies of various concepts that promised a great deal more operational flexibility and mission capability than the relatively low-speed helicopter. In addition to exploring the potential for higher cruise speeds, designers and military planners became interested in the possibility of disbursed aircraft operations from unprepared sites. In particular, the vulnerability of military bases with fixed runways to enemy attack (particularly in Europe) became a major concern that might be resolved by deployment of this new category of aircraft.

Stimulated by the possibility of military applications, the government and industry began intense studies of Vertical/Short Takeoff and Landing (V/STOL) concepts in the early 1950s. In support of the military services, the NACA and later NASA conducted valuable research on V/STOL concepts in wind tunnels, piloted simulators, propulsion test stands, and flight. Internationally recognized NASA V/STOL experts were sought

out for consultation on new programs, and NASA test pilots were invited on numerous occasions to evaluate prototype V/STOL aircraft. These V/STOL support activities have continued for more than 50 years, including efforts on today's F-35 fighter. The Langley Research Center and the Ames Research Center have led V/STOL research studies, which reached a peak in the 1960s. Most of the radical concepts proposed resulted in designs that failed to enter military service for various reasons, and only the AV-8 Harrier became operational.

Tailsitter Concept

During World War II, three technical occurrences stimulated military planners toward an interest in vertical takeoff and landing aircraft. First, the vulnerability of aircraft carriers to the Japanese kamikaze threat and airfield damage from enemy bombers suggested a need for deployment of aircraft from smaller ships or unprepared sites. Second, the initial success and potential capabilities of the helicopter demanded the continued development of vertical-takeoff airplanes with higher cruise speeds. The last stimulant was the Allied discovery of the design plans for the German Focke-Wulf Triebflugel Vertical Takeoff and Landing (VTOL) interceptor. In response to these factors, the Air Force and Navy initiated Project Hummingbird in 1947 to conduct design studies of VTOL aircraft.

The first U.S. experimentation with propeller-driven VTOL aircraft focused on the concept known as the "tailsitter" as inspired by the Focke-Wulf design. The nickname was very appropriate because the aircraft sat on its tail prior to and after flight with its nose pointing upward in a vertical attitude. Perhaps the simplest of all V/STOL concepts, key requirements for success were a relatively high thrust-to-weight ratio for vertical takeoffs and landings, and the ability to perform a controlled transition from hovering and vertical flight to conventional aircraft flight attitudes. Such designs were of particular interest to the Navy, whose intention was to deploy the aircraft on board special ships in the role of a fleet or convoy defender.

The Navy shared with Langley its interest in the tailsitter idea, thus Langley researchers conducted exploratory free-flight remotely controlled model tests of simple tailsitter configurations to evaluate vertical take-off, hovering, and vertical landing characteristics as early as 1949—in still air in a large building. The early model configurations consisted of simple cylindrical fuselage shapes, rudimentary wings and tail surfaces, and counter-rotating large-diameter propellers to minimize adverse propeller slipstream effects. Initially, NACA studies were directed at determining the adequacy of conventional aerodynamic control surfaces on the wings and tail to provide satisfactory control of aircraft motions during hovering flight and low-speed maneuvers. Since the control surfaces were located in the high-energy slipstream of the propellers, it was anticipated that control levels would be high. These experiments confirmed the control power available and produced encouraging results that further accelerated the Navy's interests in the tailsitter concept.

Convair XFY-1

The first tailsitter project, initiated by the Navy in May 1951, was the Consolidated Vultee (Convair) XFY-1 "Pogo." The aircraft configuration used two counter-rotating propellers, a delta-shaped wing, and two vertical fins. Control surfaces consisted of flap-type elevons on the wing and rudders on the vertical tails. The aircraft was nicknamed Pogo nickname because it reminded observers of a pogo stick during vertical operations. The aircraft was to be positioned on a makeshift flight deck on a surface ship, and it was to take off and land vertically. Prior to its development and first flight, however, the Pogo was the subject of extensive research carried out at NACA Langley. An earlier simplified model had been designed and tested by Langley as a preliminary study to obtain some early control information long before the scale model of the XFY-1 was constructed.

In 1951, the Navy's Bureau of Aeronautics requested that Langley researchers study the dynamic stability and control characteristics of a 0.13-scale flying model of the Pogo. These tests, which included evaluations of the model in hovering, take-off, and landing flight modes, were performed in the return passage of the full-scale

Early Vertical Takeoff and Landing (VTOL) tailsitter configuration studied at Langley in 1950. Early research focused on evaluation of stability, control, and handling qualities of rudimentary configurations in hovering flight. Tests of this remotely controlled model were conducted in a large, open building. The model exhibited an unstable oscillatory motion in pitch due to the propeller's dynamics in translating flight, but the conventional aircraft-control surfaces immersed in the high-energy prop slipstream were very powerful, and the motions could be easily controlled. (Photo courtesy of NASA)

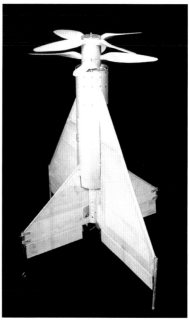

Another Langley VTOL tailsitter configuration in 1951 with contra-rotating propellers and delta wing, designed to incorporate some of the features of the evolving XFY-1 design. The model was tested for a quick look at its behavior before the XFY-1 flying model was available. Generic research models such as this one generated valuable fundamental information on stability and control requirements for hovering flight, both out- and in-ground effect. Early results obtained with this model helped Convair plan its approach to flight tests of the XFY-1 aircraft. (Photo courtesy of NASA)

tunnel. Most of the evaluations were conducted in still air conditions with the model hovering at altitude or near the ground to determine ground effects. However, some evaluations were performed with the tunnel running at simulated full-scale wind speeds approaching 40 knots. Test results revealed that the model displayed unstable pitch and yaw oscillations when controls were fixed during hovering flight, but the model's motions were slow enough to be easily controlled. Vertical take-offs and landings were made with ease. The model was generally stable during translating flight as the tunnel speed was increased, but it was hard to control in yaw when the simulated wind speed surpassed 30 knots.

The behavior of the XFY-1 during transitions from hovering to forward flight was studied in 1952 in the full-scale tunnel by flying a free-flight model at various airspeeds in the transition process, and it was discovered that the model was most unstable at angles of attack from 40- to 50-degrees. This conclusion caused a major concern over how to interpret the results during the free-

flight transition testing technique because the airspeed of the full-scale tunnel could not be rapidly increased to simulate a rapid aircraft transition from hovering to forward flight. It was argued that, theoretically, an unstable airplane might be able to quickly transition through an unstable range of angle-of-attack before the instabilities could react and result in loss of control.

Recognizing that this issue would certainly reappear for every V/STOL configuration tested in the tunnel, the Langley staff created a novel testing technique to specifically study rapid transitions to and from hovering flight. To develop this outdoor rapid-transition testing technique, Langley acquired a large crane capable of rapid rotation, and equipped it with model power systems and remote-control capability. The crane could be rotated at angular rates up to about 20 revolutions per minute and could accelerate from rest to top speed

This multiple-exposure photograph shows the transition of a free-flight model of the Convair XFY-1 Pogo flying in the full-scale tunnel in 1952. The sequence shows the model on the tunnel floor before take-off, hovering vertically in still air, and flying forward in conventional flight as the tunnel airspeed was increased. During the test program, researchers removed the bottom vertical tail to evaluate its contribution to stability and to determine if the pilot could make a belly-landing in an emergency. The numerical index at the top of the open-jet tunnel is an airspeed indicator. The flight tests were squeezed into the full-scale tunnel's busy schedule to permit visitors from Convair (including the XFY-1 test pilot) to observe the model. Note the test set up at the lower left, which was a test stand for a full-scale rotor test. (Photo courtesy of NASA)

Langley researchers developed a unique testing technique to evaluate stability and control during rapid transitions from hovering flight to forward flight and back to hovering. The Langley Control Line Facility was developed based on a huge rapid-rotation crane that included positions for the flight crew in its enclosure. Transition tests required the entire crew to be seated in the crane. Shown here is the Convair XFY-1 free-flight model in hovering flight with pilot Robert H. Kirby seated outside the crane as other members of the test crew observe. The objective of this particular test was to determine if roll disturbances that had been experienced on the full-scale aircraft during hovering flight in an airship hangar at Moffett Field, California, would also be present in outside "free-air" conditions. Engineers thought disturbances were caused by propeller slipstream recirculation effects within enclosed structures. The full-scale Langley model results on the crane facility showed that the disturbances were not present, subsequently verified when the full-scale XFY-1 was flown outdoors. (Photo courtesy of NASA)

in about one-quarter of a revolution. Seated in the enlarged crane cab were the model-controls' operator, the safety-cable operator, the model-power operator, and the crane operator. The powered model being tested was mounted about 50 feet from the center of rotation, and was restrained by wires from the model to the crane cab to oppose the centrifugal force during rotating flight around the circle. The tests were limited to studies of the longitudinal pitching motions of the model during the transition process.

The tests began with the model in stabilized hovering flight about 15 feet above the ground; the pilot applied pitch control to begin the transition process; and the model power operator adjusted the power to ensure a constant-altitude maneuver. With this approach, the flight-test crew evolved a carefully orchestrated procedure in which thrust and control inputs were synchronized with the rotation speed of the crane to permit rapid transitions from hovering to cruise and back to hovering. The procedure resembled the control-line technique traditionally used by model airplane enthusiasts, and the facility subsequently became known as the control-line facility. Transitions could be made much more rapidly than those in the full-scale tunnel and at rates more representative of full-scale VTOL airplanes.

A 0.13-scale powered model of the XFY-1 Pogo was tested on the facility at Langley, permitting the evaluation pilots to observe the beneficial effects of rapid transitions through areas of instability. The test program included essentially constant-altitude transitions from hovering to

The Convair XFY-1 Pogo in vertical flight. Despite a successful flight program of transitions to and from hovering flight, the concept was dismissed due to operational concerns. (Photo courtesy of Smithsonian Institution via NASA)

conventional forward flight and return to hovering. The experience gained from the Pogo tests were a valuable foundation for the Langley staff in tests of other V/STOL model aircraft.

After many captive indoor hovering tests using a special tethered-flight rig in an airship hangar at Moffett Naval Air Station, California (adjacent to the Ames Aeronautical Laboratory), the full-scale XFY-1 aircraft flew outdoors successfully for the first time on June 2, 1954, and in November it completed a transition from vertical takeoff to horizontal flight and back to hovering followed by a vertical landing. Despite the Pogo's record-setting flight as the first non-helicopter aircraft to complete a VTOL mission, other major operational dilemmas were encountered, including engine and propeller problems and aircraft servicing difficulty. In addition, test pilots complained about the lack of adequate vision due to the vertical attitude of the aircraft (especially during attempts at precision vertical landings looking back over their shoulders) and the relative placement of the cockpit on the fuselage. These problems, together with the progression of the jet age and increasing reliance on jet fighters, resulted in the cancellation of the Pogo program in 1956.

The relationships and working arrangements between the personnel of Langley and Convair during the XFY-1 program were especially productive and valuable. In return for results obtained in Langley's unique facilities, Convair provided almost constant feedback on flight data and experiences that proved to be of tremendous value to the NACA staff for the validation of Langley's testing techniques and planning future V/STOL research.

Lockheed XFV-1

Along with the XFY-1 program, in June 1951 the Navy pursued the development of the Lockheed XFV-1 Salmon, another tailsitter that used contra-rotating propellers. Like the Pogo, the Salmon (named for Lockheed Chief Test Pilot Herman "Fish" Salmon) was to serve as a fleet or convoy defender. The XFV-1 featured straight wings and four vertical tail fins arranged in a cruciform X layout. Control surfaces on the tails provided control for hovering and forward flight.

Early in the program, Lockheed arranged for hovering tests of a large 0.25-scale free-flight model in the 40-by-80-foot tunnel at the NACA Ames facility. More extensive free-flight model tests of a 0.13-scale model of the XFV-1 were conducted by Langley in the full-scale tunnel prior to the aircraft's maiden flight. The study was broad in scope and included general research objectives of interest to the NACA. For example, a second research model was tested as the basic XFV-1 with an unswept wing and X-cruciform tails, but it was also modified to include variables of interest such as a swept wing and cruciform tails in a cross-orientation rather than X-orientation. Testing of the XFV-1 model included evaluations of flying characteristics for relatively high-speed sideways flight. Interestingly, the flight studies indicated no major differences between the model configurations. In hovering flight, both configurations displayed the unstable pitching and yawing oscillations that had been observed earlier for the XFY-1 Pogo model, but again movement was slow enough to permit pilots to easily eliminate the motions and precisely control the hover. During the transition to forward flight, both configurations were controllable, and for the sideward-flight studies the model was flown up to a sideward full-scale speed of about 10 knots.

Flight testing of the full-scale XFV-1 aircraft ultimately proved to be less spectacular than the XFY-1. The Navy had supplied only one engine (YT-40) rated for vertical take offs, and Convair had been given that powerplant. The XFV-1 used a lower powered engine. As a result, the aircraft only performed horizontal takeoffs and landings, and it required a special fixed horizontal-attitude landing gear. Although more than 30

The Lockheed XFV-1 tailsitter. Although the design was able to transition to and from hovering flight at altitude, it never achieved a vertical takeoff to hover because of engine limitations. A special ladder was required for pilot entry and egress, and a cradle was used to rotate the aircraft to a vertical position. (Photo courtesy of Lockheed Martin via Eric Hehs)

Rear view of the XFV-1 showing its cruciform "X" tail arrangement. Note the tail wheels on the fin tips and the ungainly fixed landing gear used in the flight program. (Photo courtesy of Lockheed Martin via Eric Hehs)

Powered free-flight models of the Convair XFY-1 (left) and the Lockheed XFV-1 (right) were tested in the Langley Full Scale Tunnel. In the case of the XFV-1, a second research model was also tested with the baseline wing as well as an alternate swept wing with little difference noted in characteristics from the baseline design. Langley researcher Robert O. Schade peers from one of the pilot's positions. (Photo courtesy of NASA)

The XFV-1 in conventional flight above Edwards AFB, California. The conventional landing gear was incorporated in the design for horizontal take-offs and landings. (Photo courtesy of Lockheed Martin via Eric Hehs)

transitions were made at altitude from cruising to hovering flight, no vertical takeoffs or landings were made. The XFV-1 program came to an abrupt end with its cancellation in 1956 due to the same shortcomings as the Pogo program.

Ryan X-13

In 1953, the Air Force pursued the development of a VTOL turbojet tailsitter design based on extensive VTOL development testing undertaken by Ryan for a Navy contract. Ryan had been conducting general studies of jet-reaction controls under the Navy contract, including studies of a flying test rig consisting of a remote-controlled vertical-attitude engine. The Ryan turbojet aircraft design, which became known as the X-13, was tiny (wing span of 21 feet and length of 24 feet) and featured a 10,000-pound-thrust Rolls-Royce Avon turbojet engine for propulsion. It was equipped with a swiveling engine tailpipe to provide pitch and yaw control, and wingtip air nozzles supplied with compressed air from the engine provided roll control. The aircraft was designed to be launched and recovered from a special trailer oriented in a vertical position. A nose hook was mounted on the X-13 and used to engage a "clothesline" wire for vertical takeoffs and landings.

Early in the X-13 program, Langley conducted hovering and transition flight tests of a 0.2-scale model of the aircraft to determine its dynamic-stability behavior. Tests were conducted in both the full-scale tunnel and the control-line facility beginning in 1954. In order to produce enough thrust required for vertical flight for this jet configuration, researchers used a compressed-air jet exhausting through an injector tube for the indoor wind-tunnel flight tests. For the outdoor control-line tests, a more hazardous high-energy hydrogen-peroxide rocket motor was used. The model replicated the critical features of the full-scale airplane, including a swiveling engine exhaust nozzle and roll-control jets on each wingtip. In addition, an artificial stabilization system in pitch and yaw was used to simulate the control system feedbacks used for the full-scale aircraft.

The test program included an investigation of the effects of gyroscopic cross-coupling caused by the gyroscopic precessional moments created by the relatively large jet engine during angular movements. Because of the relatively small size of the X-13 airframe relative to the engine, Langley researchers anticipated that the gyroscopic moments would be quite large and would affect controllability during transitions. Therefore, they equipped the model with a large rotating flywheel driven by an electric motor to simulate the engine's gyroscopic effects. Flight-test results showed that, in hovering flight, the model could be flown smoothly without any automatic stabilization when the jet engine gyroscopic effects were not simulated. When

Langley technician Joe Block performs maintenance on the free-flight model of the Ryan X-13 Vertijet. The model was used for both indoor and outdoor flight tests. In the Langley Full Scale Tunnel it was powered by compressed air, whereas a hydrogen-peroxide rocket motor was used during tests on the Langley Control Line Facility. (Photo courtesy of NASA)

Hydrogen-peroxide rocket-powered free-flight model of the X-13 in rapid-transition flight evaluations on the control line facility. During these tests the flight crew (pilot, thrust controller, and crane operator) were seated in the crane's cabin. Prolonged flights using the revolving crane were almost guaranteed to induce motion sickness on the test crew. (Photo courtesy of NASA)

the gyroscopic effects were simulated, the model could not be controlled in hovering flight without artificial stabilization because of strong coupling of pitching and yawing motions. With artificial stabilization, the model motions were heavily damped and the model was easy to control. In transition flight, artificial stabilization was always used and the model was easy to fly, even with the jet engine gyroscopic effects present. The swiveling engine nozzle provided powerful yaw and pitch control in all flights.

Initial flight testing of the full-scale X-13 began in 1956, and the aircraft performed test flights starting in the vertical position from a special flat-bed trailer for the first time in 1957. The aircraft completed more than 100 successful flight tests in which its VTOL capability was dramatically demonstrated. During one spectacular demonstration, the aircraft took off from its vertical trailer in front of the Pentagon in Washington, D.C., performed a transition to forward flight in front of observers, and then performed a transition back to hovering flight and performed a vertical landing on its

trailer. The Air Force, however, agreed with the Navy's concern over aircraft maintenance and piloting difficulties for vertical attitude VTOL configurations. Consequently, the X-13 program was cancelled. After experiences with the XFY-1, XFV-1, and X-13, the services concluded that acceptable VTOL designs would require a horizontal attitude for vertical takeoff and landings. The vertical-attitude concept then fell into oblivion until the mid-1970s, when Langley researchers investigated an even more radical approach.

"Woodpeckers"

As frequently happens in research and technology, new advances in aircraft technology in the 1970s stimulated a new examination of the vertical-attitude concept. Researchers were aware that three primary factors caused the rejection of vertical-attitude VTOL aircraft in the 1950s: (1) the high level of thrust required for vertical takeoff and landing was much greater than that required for cruising flight, resulting in

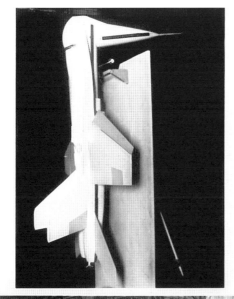

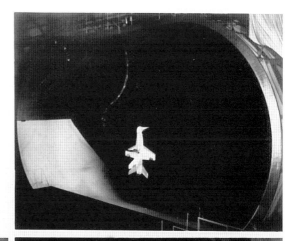

The Langley-conceived "Woodpecker" VTOL version of the YF-17 lightweight fighter design tested in the Langley Full Scale Tunnel. Advances in technology of the 1970s suggested apparent solutions to past problems of vertical-attitude VTOL aircraft. Sequence shows the Woodpecker free-flight model prior to takeoff, mounted on a platform similar to that used by the X-13, followed by a vertical takeoff to hovering flight and transition to conventional flight as speed was increased in the full-scale tunnel. Note the pivoting cockpit, which was maintained in a conventional attitude during flights. (Photo courtesy of NASA)

excessive engine size and weight, (2) requirements for special ground-based takeoff and landing platforms were unacceptable; and (3) the vertical attitude of the cockpit during takeoff and landings was objectionable to pilots during operations, especially during landing.

However, in the early 1970s a new generation of advanced-technology lightweight fighters known as the YF-16 and YF-17 incorporated features that could possibly minimize or even eliminate some of the shortcomings of the earlier tailsitters. These new aircraft had uninstalled thrust-to-weight ratios of almost 1.5 in order to meet the high transonic performance requirements of the time, and they used fly-by-wire flight-control systems that would potentially eliminate the need for mechanical control linkages between the pilot and the control surfaces. The Langley researchers reasoned that by adding a jet-reaction control system for control while hovering in low-speed flight, a landing hook on the lower fuselage for engaging a landing wire, and a pivoted-nose-cockpit arrangement that permitted the pilot to remain in normal horizontal position during operations, an acceptable VTOL fighter for Navy operations at sea might be feasible. Interest in the new VTOL concept, nicknamed the "Woodpecker," stimulated tests at Langley.

At the request of the Air Force, Langley researchers had supported the earlier development programs for the conventional YF-16 and YF-17, including free-flight tests in the full-scale tunnel to evaluate high angle-of-attack characteristics. The models used for these studies

Free-flight model of the Woodpecker version of the YF-16 during transition flight in the full-scale tunnel. The model was tested with its basic swept wing and with a modified delta wing. High relative humidity resulted in condensation and visualization of the compressed air-thrust stream. (Photo courtesy of NASA)

were made available by the Air Force for the NASA VTOL research, and were modified for the exploratory tests. The models were equipped with a remotely controlled pivoting fuselage forebody that could be rotated 90 degrees during the transition to and from vertical flight so as to maintain a conventional cockpit attitude relative to the horizon. Longitudinal controls consisted of a conventional all-moving horizontal tail and a jet-reaction controller at the rear of the fuselage. Lateral and directional controls included aileron and rudder surfaces as well as jet-reaction controls at the wingtips and tail for roll and yaw control. The YF-16 model was tested with its basic swept wing and a delta wing. The additional wing shape was selected because the delta wing was expected to provide better aerodynamic characteristics during the transition to forward flight. The flights began with the tunnel off. In the still air, the model made a vertical take-off from the landing platform, transitioned from hovering to forward flight with the cockpit essentially horizontal, and ended with the model in conventional forward flight at high angles-of-attack. The tests included steady flights at several attitudes, an examination of the control power used, and an evaluation of the need for artificial rate damping.

Flight tests concluded that the YF-16 model (with both wing shapes) and the YF-17 model exhibited satisfactory longitudinal stability and control characteristics; however, without artificial stabilization, all the configurations were very difficult to control during transitions, and the delta-wing version of the YF-16 exhibited very large rolling and yawing motions. With artificial rate dampers operating, very smooth transitions could be made consistently from hovering to normal forward flight with relatively little difficulty. The results of these radical tests were briefed to the Pentagon and discussed with and transmitted to industry and the Navy, but no follow-on activities occurred.

Tilt-Wing Concept

Also during the 1950s, NACA researchers at Langley under Charles Zimmerman (who had returned from Vought) began to experiment with propeller-driven V/STOL designs that used the tilt-wing concept. In this concept, the aircraft wing and propellers are tilted from a vertical attitude relative to the fuselage for hovering flight to a horizontal position for cruise flight. The tilt-wing concept has been a very strong candidate for V/STOL operations, and it continues to be studied to the present day for applications to military transports. Several U.S. tilt-wing research aircraft have been developed and flown, including the Vertol VZ-2, the Hiller X-18, and the XC-142A. Each of these aircraft programs was strongly supported by staff at Langley.

During early research efforts, it was quickly learned that the most critical problem inherent in the tilt-wing concept is the tendency of the wing to experience large areas of flow separation and stalling at the high angles of incidence required during the transition to and from

Langley technician Clarence Breen prepares a multi-engine tilt-wing free-flight model for flight testing in the full-scale tunnel in the early 1950s. Note the control linkages for the variable-pitch propellers that provided roll control at low speeds, and the jet-reaction control box at the tail used for pitch control at low speeds. Charles Zimmerman was the Langley advocate for these efforts. (Photo courtesy of NASA)

hovering to forward flight. One of the primary factors that prevent the wing from stalling is the immersion of the wing in the high-energy slipstream of the propellers. Therefore, the probability of wing stall is significantly higher during steeper descents, when power is reduced and the slipstream energy is diminished. When wing stall occurs, tilt-wing aircraft usually exhibit random, uncontrollable rolling, wing-dropping motions and/or yawing motions, and generally unsatisfactory flight characteristics. To minimize and prevent this problem, the geometry of wing flaps and leading-edge stall control devices must be carefully selected.

Vertol VZ-2

Following very successful pioneering research by Langley using generic propeller-driven free-flight models, the Army Transportation Corps and the Office of Naval Research collaborated to develop and build the tilt-wing VZ-2 research aircraft to investigate the practicality of the tilt-wing concept. The project was designed around a low-cost flight research aircraft manufactured by Vertol. A single turbine engine transmitted power by mechanical shafting to a pair of three-blade propellers and to two tail-control fans, which provided yaw and pitch control in hovering flight. Roll control was provided by differential thrust between the two vertically oriented propellers in hovering flight.

The VZ-2 flew for the first time in 1957. It successfully completed the world's first transition from vertical to horizontal flight by a tilt-wing airplane on July 15,

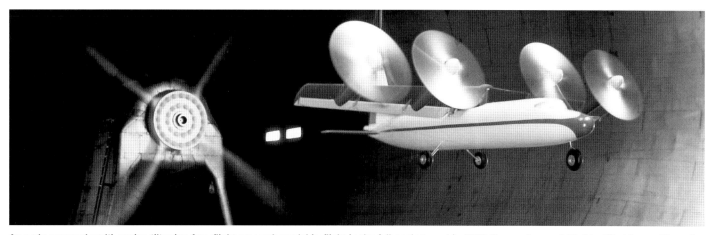

An early powered multi-engine tilt-wing free-flight research model in flight in the full-scale tunnel in 1955. Researchers quickly identified the problem of wing stall during transition as being a major barrier to this concept, requiring careful design of wing leading- and trailing-edge devices to prevent unacceptable handling qualities, particularly during descending flight at partial power. Uncontrollable wing-dropping motions encountered when the wing stalled severely limited the usable descent rate of a poorly designed configuration. (Photo courtesy of NASA)

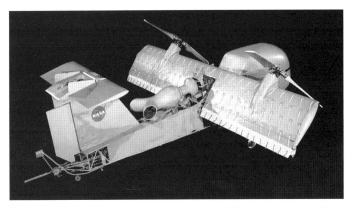

Free-flight model of the Vertol VZ-2 tested in the full-scale tunnel at NASA Langley. Note the articulated propeller blades and the jet thruster mounted at the rear end of the fuselage for pitch and yaw control in hovering flight. Mechanization of the tail rotor used for control on the full-scale aircraft was deemed too complicated to be implemented on the model. (Photo courtesy of NASA)

The VZ-2 free-flight model in hovering flight before transitioning to forward flight in the full-scale tunnel. The anticipated wing-stall issues and deterioration in handling qualities for such conditions were encountered in the model flight tests. Wing leading-edge modifications significantly improved flying characteristics. (Photo courtesy of NASA)

Rear view of the VZ-2 mounted in the full-scale tunnel for aerodynamic testing in 1961. Extensive studies of the wing-stall phenomenon were conducted in the tunnel, and results were correlated with flight measurements. (Photo courtesy of NASA)

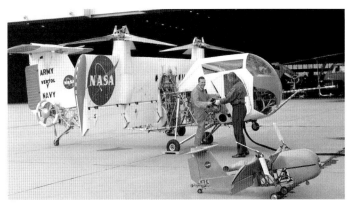

NASA Langley test pilot Robert Champine (left) poses for a publicity photo in front of the Langley flight hangar with the Vertol VZ-2 tilt-wing flight research aircraft and the VZ-2 wind-tunnel free-flight model that was tested in the full-scale tunnel. The VZ-2 project was characterized by closely coordinated wind-tunnel and flight activities within Langley, industry, and DoD. (Photo courtesy of NASA)

1958, at Edwards Air Force Base. In 1959, NASA Langley received the VZ-2 from Edwards for extensive in-house flight-research experiments. Interestingly, the Army did not have pilots who were qualified for V/STOL aircraft, so Langley pilots conducted the research missions of interest to the Army.

The extent of VZ-2 supporting research at Langley included sub-scale and full-scale wind-tunnel tests, free-flight model tests, and flight evaluations. During the VZ-2 flight experiments at Langley, research pilots cited problems with poor low-speed stability and control, and deficient handling qualities during transition. They encountered the anticipated wing-stall problem in the speed range from about 40- to 70-knots, which corresponded to wing incidence angles from about 45 degrees down to 25 degrees. In its original configuration, the airplane had no wing flaps or leading-edge high-lift devices and exhibited severe wing stall, heavy structural buffeting, and random rolling and yawing motions, especially in descending flight. The situation in descending flight was so unsatisfactory that pilots considered it an area of hazardous operations. To solve the problem, researchers recommended the use of a wing-section modification consisting of a drooped wing leading edge and an increase in leading-edge radius. This modification alleviated much of the wing stall and the unsatisfactory lateral-directional motions and permitted a significant increase in the useable rate of descent.

Following the initial Langley flight tests, the full-scale VZ-2 was mounted in the full-scale tunnel in 1961

The Vertol VZ-2 tilt-wing flight research aircraft mounted in the full-scale tunnel for aerodynamic testing following initial flight-test evaluations by NASA pilots. Note that the aircraft had ailerons, but no trailing-edge flaps at the time of the test. The test program included assessments of the benefits of leading-edge slats and leading-edge droop. (Photo courtesy of NASA)

for wind-tunnel/flight correlation studies with emphasis on the wing-stall phenomenon. Additional analysis indicated that the use of a large trailing-edge flap would significantly augment the lift required during transition, thereby permitting the wing to operate at lower angles-of-attack where stalling was less of a problem. The VZ-2 was subsequently modified with wing changes, including a full-span trailing-edge flap. In hovering flight, the aircraft had poor control harmony. The level of control in hovering was inadequate in yaw, marginal in pitch, and excessive in roll. Langley pilots also recommended that an artificial rate stabilization system be installed to alleviate erratic aircraft motions caused by ground effect encountered when the hovering height decreased to less than 19 feet.

The flight-test programs for the basic and modified VZ-2 proved to be successes, with 34 conversions between vertical and horizontal flight. The program lasted until 1964, and the aircraft, which was designed as a rudimentary research vehicle with no intentions of production, was later donated to the Smithsonian Institution. Besides its pioneering research for the tilt-wing V/STOL concept, the program resulted in invaluable experience and guidance for future programs such as the Hiller X-18 and the LTV-Hiller-Ryan XC-142A tilt-wing transports.

Hiller X-18

In 1956, military interest in the tilt-wing V/STOL concept had intensified, and the Air Force awarded a study contract to the Hiller Aircraft Corp. for the design and flight testing of a low-cost transport concept designated

the X-18. The twin-engine X-18 was designed to be a low-cost test bed to explore the feasibility of a large tilt-wing transport aircraft. It used three-blade, counter-rotating propellers and turboprops from the cancelled Convair XFY-1 and Lockheed XFV-1 programs and a Chase C-122 fuselage (originally a Chase XCG-18 glider). With a maximum gross weight of 33,000 pounds, the X-18 was a second-generation development of tilt-wing technology more representative of potential military transport configurations. The X-18 weighed almost 10 times more than the VZ-2. The engines of the X-18, however, were not connected by a cross-shaft arrangement, posing a safety-of-flight concern if an engine stopped during vertical flight. Pitch control for hovering and low-speed flight was achieved by vectoring the thrust of an internally mounted Westinghouse J34 turbojet through a tailpipe, while roll control was produced by differential engine thrust and yaw control by ailerons in the propeller slipstreams. In conventional flight, the X-18 used ailerons, elevator, and rudder controls. Wind-tunnel testing began in 1957, and the first flight of the full-scale aircraft occurred at Edwards Air Force Base in 1959. The flight test program took place from 1960 to 1961.

In support of the Air Force program, Langley conducted conventional wind-tunnel aerodynamic tests and free-flight tests of a relatively large (six-foot wingspan), 0.125-scale model of the X-18 in the Langley Full-Scale Tunnel in 1957 and 1958. Aerodynamic characteristics were measured prior to the model flight tests, with particular emphasis on wing-stall behavior. The flight tests consisted of slow constant-altitude transitions from hovering flight to forward flight, hovering tests near the ground, and vertical takeoffs and landings.

Langley technician Joe Denn prepares a large powered free-flight model of the Hiller X-18 tilt-wing experimental aircraft for tests to measure its aerodynamic characteristics in the full-scale tunnel in 1957. The model used dual counter-rotating propellers to simulate the full-scale aircraft propulsion system, and was powered for these tests by electric motors with power routed up the model test apparatus. So-called "force tests" using internally mounted electric strain-gauge balances are routinely conducted to document and analyze aerodynamics before free-flight tests are started. (Photo courtesy of NASA)

Powered free-flight model of the X-18 midway through a transition to forward flight in the full-scale tunnel. The cylindrical unit under the aft fuselage replicates the pipe used by the full-scale aircraft to transmit the exhaust of an internal turbojet engine to the tail for pitch control. For the model in hovering flight, ailerons were used for yaw control, a compressed-air jet at the tail provided pitch control, and roll control was produced by differential pitch of the rear elements of the dual propellers. (Photo courtesy of NASA)

A few flights were made to evaluate characteristics in rearward and sideward flight.

The Langley researchers found that the X-18 model exhibited unstable, divergent oscillations in roll and pitch when the controls were fixed during hover. This same characteristic had been observed in earlier hover testing of generic tilt-wing configurations and in the VZ-2 model as previously discussed. The unstable motions were relatively slow to build up, with periods on the order of five seconds at model scale, and were therefore easy to control. The levels of pitch control projected for the full-scale aircraft in hover were expected to be very satisfactory, but the level of roll control was insufficient, and the variable-pitch propellers required larger deflections for adequate control. The tests also revealed a potential problem with center-of-gravity location during transition to forward flight, and provided guidance for the full-scale design.

In its 20-flight test program at Edwards, the full-scale X-18 achieved a maximum wing tilt angle (relative to the fuselage) of 33 degrees and never achieved hovering flight. It was found to be a stable, docile aircraft in conventional flight. Most of the flight tests were conducted at altitude following a conventional takeoff. Although severe wing buffeting was experienced, as in other tilt-wing projects, a

drooped-wing leading edge reduced the severity of the problem. On the 20th and last flight on March 4, 1960, a propeller pitch failure occurred on one engine with the wing tilted about 25 degrees and the aircraft at an altitude of about 7,000 feet. The resulting aerodynamic asymmetry caused the aircraft to quickly roll into an inverted spin, followed by frantic efforts of the aircrew to recover. Recovery was successfully completed at a very low altitude, and the X-18 never flew again.

Following the termination of the flight-test program in July 1960, the X-18 was used for ground-effects testing using a special test stand at Edwards. The project's objective was to measure and collect data on ground effects and slipstream recirculation effects, which were favorable for the X-18. The test stand failed during tests in late 1963, leading to a termination of the X-18 program. Although the X-18 program was short lived and never explored the V/STOL envelope in depth, it produced a wealth of data for use in the subsequent program for the XC-142A transport.

Ling-Temco-Vought XC-142A

At the end of the 1950s, the Army, Air Force, and Navy had all continued to stimulate V/STOL developments, joining together for a Tri-Service V/STOL transport project with a request for proposals sent to industry teams in 1961. Competing models were narrowed to a four-engine, tilt-wing concept designed by   a team of Ling-Temco-Vought, Ryan, and Hiller; and a contract for five aircraft was awarded. The XC-142A was a large (maximum gross weight of 41,500 pounds), fast (top speed of 400 mph), tilt-wing design with extensive

cross-shafting of main engines for safety and a tail-mounted, 3-blade, variable-pitch propeller for pitch control at low speeds.

In response to military requests, Langley conducted conventional wind-tunnel model tests in its full-scale tunnel and 300-mph 7- x 10-foot tunnel to analyze the configuration's V/STOL characteristics. These static tests were augmented by free-flight model tests in the full-scale tunnel in 1964 using a large 0.11-scale model. Spin-tunnel tests of an unpowered XC-142A model were also conducted. In hovering flight, roll control for the XC-142A was provided by differential deflections of the propeller blade pitch of

The large 0.11-scale free-flight model of the XC-142 multi-engine tilt-wing experimental research aircraft studied in the full-scale tunnel. Previous research on the VZ-2 and X-18 had alerted designers to the wing-stall problem for tilt-wing configurations, and the XC-142 used leading- and trailing-edge wing devices to minimize the issue. The all-movable horizontal tail was interconnected to the wing-tilt mechanism so as to avoid large power-induced trim changes during transition flight. (Photo courtesy of NASA)

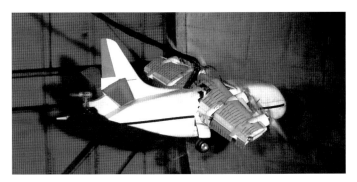

The XC-142 free-flight model in flight in the full-scale tunnel in 1964. Note the wool tufts on the upper wing surface for flow visualization of wing stall during transition flights. The flight tests were uneventful, and the model's characteristics were satisfactory. (Photo courtesy of NASA)

One of two XV-6 Kestrel aircraft flown in research at NASA Langley and the Ling-Temco-Vought XC-142 multi-engine tilt-wing experimental research aircraft pose in front of the Langley flight research hangar in 1968. (Photo courtesy of NASA)

The XC-142 in hovering flight at Langley during NASA evaluations. The flight program focused on general research for potential applications of the tilt-wing concept to future aircraft, and to civil V/STOL operations in the terminal area of airports. Noise signatures of the XC-142 were also measured and compared to predictions. (Photo courtesy of NASA)

the four main propellers, yaw control was provided by differentially deflecting the ailerons, and pitch control was provided by varying the blade angle of the tail-mounted rotor. These control functions were phased out as the transition to forward flight progressed, resulting in the use of conventional control surfaces.

Free-flight results showed the existence of unstable control-fixed oscillations in pitch and yaw in hovering flight, similar to results obtained for previous model tests of the VZ-2 and X-18 tilt-wing configurations. Once again, the unstable oscillations were so slow (a period of about 10 seconds for the full-scale airplane) that they could be easily controlled and the model could be smoothly maneuvered, even without artificial stabilization. No problems were noted during transitions in level flight or in simulated descents, and the minimum control power found to be satisfactory for the model was less than the control power planned for the full-scale aircraft.

The five XC-142A aircraft underwent military evaluations from 1964 to 1967, and four of the aircraft were severely damaged in incidents or lost in crashes. The flight-test program demonstrated that serious stability

and control issues remained with the design, particularly during operations close to the ground or in descending flight. At the end of the planned flight-test program, the services could not define an operational requirement for a V/STOL transport, and the last XC-142A aircraft was loaned to NASA Langley for general V/STOL flight research studies from 1968 to 1970.

The XC-142A flight-research program at NASA Langley, which began in October 1968, involved a broad scope of investigations, covering topics that ranged from assessments of V/STOL performance and handling qualities to research on civil V/STOL terminal-area operations in instrument flight conditions. In addition, special tests

One of the issues uncovered in the tilt-wing flight programs of the VZ-2, X-18, and XC-142 was the undesirable effects of turbulent recirculation of the propeller slipstream in ground effect. The impact of the recirculating flow caused aircraft buffeting and wallowing motions near the ground. Here the XC-142 is approaching for a short landing with smoke used to visualize the flow in front of the aircraft. Note the areas of recirculating flow billowing back into the descending aircraft. (Photo courtesy of NASA)

Time-lapse photo sequence shows the successful transition of the Doak VZ-4 tilt-duct testbed from hovering to conventional flight. NASA test pilot Jack Reeder escaped from the aircraft following an engine fire during the flight program at Langley. (Photo courtesy of NASA via Jack Reeder)

were conducted to investigate the phenomenon of airframe buffeting and "wallowing" aircraft motions caused by ground-induced turbulent recirculation of the propeller/wing slipstream during STOL operations. Noise levels created by the XC-142A in hovering and conventional landing approaches were also measured to generate data for guidance in civil applications of V/STOL aircraft, as well as for validation of propeller noise-prediction methods. Flight activities were conducted at Langley and from the NASA Wallops Island, Virginia, flight-test facility. Despite the stability and control problems mentioned earlier, the XC-142A flight-test program was successful, showing the V/STOL potential of a large tilt-wing propeller-driven transport. At the project's conclusion, Langley pilots delivered the XC-142A to the Air Force Museum at Dayton, Ohio, in May 1970.

By the early 1970s, during the Vietnam War, the U.S. military and its leaders showed no further interest in developing V/STOL aircraft other than the helicopter. With no potential mission or interest, tilt-wing V/STOL aircraft research and development was terminated. The tilt-wing V/STOL concept, however, is still regarded as having high potential for military and civil applications and resurfaces from time to time in proposals today.

Tilt-Duct Concept

Another V/STOL concept explored by researchers at NASA Langley was the tilt-duct concept, which uses propellers or fans enclosed in tilting ducts that can be rotated between vertical and horizontal attitudes relative to the airframe. The tilt-duct design was pursued because the shrouded propeller offered the promise of benefits such as enhanced thrust for a given propeller diameter (resulting in a more compact design), improved operational safety for air and ground crews because of the protective shroud, and aircraft noise alleviation.

Doak VZ-4

In 1956, when the U.S. Army pursued the development of a tilt-duct research test bed, an aircraft designated the VZ-4 was contracted from the Doak Aircraft Co. The VZ-4 used a Lycoming T53 engine to power two wingtip-mounted ducted propellers via a transmission system for V/STOL operations. Pitch and yaw control in hovering flight were provided by deflecting vanes in the turbine exhaust at the tail of the aircraft, and roll control was provided by differential thrust of the ducted propellers, accomplished by varying the angle of radially arranged inlet guide vanes in each duct. The VZ-4's initial flight tests were made by Doak in 1958, and the vehicle was subsequently accepted by the Army and assigned to NASA Langley in early 1960 for flight-research studies.

During flight tests of the VZ-4 at Langley, research pilots found that the aircraft had an undesirable habit of nosing up during the transitional flight phase from

The Doak VZ-4 tilt-duct testbed at NASA Langley in 1960. The VZ-4 was found to exhibit undesirable nose-up tendencies during the transition from hover to forward flight and relatively poor STOL performance compared to other concepts. (Photo courtesy of NASA)

Bell X-22A

In 1961, Bell Aerospace had entered the tri-service V/STOL transport competition with a tilt-duct design based on in-house research and development of several tilt-duct layouts. After losing to the XC-142A competition, Bell continued its interest in tilt-duct design and subsequently was awarded a Navy contract to build and flight-test two X-22A tilt-duct research aircraft in 1962. The first aircraft was rolled off the assembly line in 1965 and first hovered in March 1966. This aircraft was stricken from the inventory shortly thereafter, following dual hydraulic failures that caused a non-fatal crash landing on August 8, 1966. Bell conducted thorough flight tests of the second X-22A until 1971, when the Navy acquired it for further testing.

The X-22A was powered by four turboshaft engines that turned four three-blade ducted propellers, driven through a power-transmission system between the engines and propellers. Control surfaces consisted of four vanes located in the slipstream of each of the ducted propellers. In forward flight, pitch control was provided by differential deflection of the front and rear vanes, while roll control was produced by differential deflections of the right and left side vanes. Differential propeller thrust maintained yaw control in cruise, and in hovering flight, roll control was produced by differential propeller thrust, and pitch and yaw control were provided by vane deflections. During the transition to and from

hover to cruise flight. Analysis attributed the problem to the ducts' design, and supplemental wind-tunnel studies indicated that placing vanes in the duct exhausts and adjusting the pivot points of the ducts could alleviate the problem. This suspicion was confirmed in wind-tunnel tests, and the VZ-4's ducts were modified with the addition of control vanes.

Another deficiency noted in the evaluation was the VZ-4's relatively poor STOL performance compared to other V/STOL concepts, such as the tilt-wing configuration. This shortcoming is an inherent feature of tilt-duct designs, caused by an inefficient spanwise lift loading during short-takeoff operations. Finally, the aircraft's useful rates of descent were limited because of stalling of the outer, fixed-wing panels, which resulted in buffeting and erratic wing dropping and wallowing motions. One flight test nearly resulted in tragedy when the VZ-4's engine caught on fire. Langley test pilot Jack Reeder was plucked from the fire by ground crews before the vehicle was fully engulfed. The VZ-4 was transferred to the U.S. Army Transportation Command Museum at Ft. Eustis, Virginia, in August 1972.

Despite the aforementioned problems, the VZ-4 provided significant information on the characteristics of tilt-duct V/STOL designs. These results, together with an interest in the compactness of the tilt-duct configuration for shipboard operations and industry activities in research on tilt-duct designs, directly influenced the Navy into pursuing the development of a more advanced, four-engine, tilt-duct Bell X-22A.

Project engineer William A. Newsom prepares a 0.18-scale free-flight model of the Bell X-22 tilt-duct experimental research aircraft for tests in the full-scale tunnel. The inlet radius of the ducted fans on the model had to be increased after this photograph was taken to more accurately simulate the aerodynamic behavior of the full-scale units. (Photo courtesy of NASA)

forward flight, the control inputs were mixed as a function of the duct angle.

Prior to the Bell and Navy flight tests, extensive NASA wind-tunnel testing of the X-22A took place at Langley and at Ames. At Langley, free-flight, spin-tunnel, and conventional models of the X-22A were tested. Spin-tunnel testing of a 0.05-scale model was started in January 1964 to provide guidance for spin recovery for the radical X-22 configuration. Later that year, a 0.18-scale free-flight model of the X-22A was tested and flown in the full-scale tunnel. The free-flight model incorporated control-system surfaces similar to the full-scale aircraft; however, it was not feasible to replicate the differential propeller pitch mechanisms, and reaction jets powered by compressed air were substituted. The model was also designed with a larger, unscaled inlet lip radius on the propeller ducts because earlier Langley research in other wind tunnels had shown that a scaled inlet lip radius on sub-scale models could not simulate the flow of air into the inlet without the inlet lip stalling prematurely. With this modification, the model's aerodynamic behavior more closely represented that of the aircraft.

The free-flight results showed the usual unstable oscillations in pitch and roll that had occurred with controls fixed for other propeller-driven V/STOL designs in hovering flight, and the period of oscillatory motion (about eight seconds full-scale) was also slow and easy to control. The minimum control power level required for satisfactory behavior in all flight modes was found to be equal to or less than half that proposed for the full-scale aircraft. The powerful automatic stabilization system and variable-stability features of the full-scale X-22A completely masked the few issues observed in the model flight tests.

The X-22A flight-test programs for both Bell and the Navy later focused on providing a capability for in-flight simulation of V/STOL aircraft characteristics, which would be used for research on handling qualities and pilot training. This part of the X-22 program met with great success, and research flights of the X-22A by the Calspan Corp. of Buffalo, New York, continued until 1984. However, the Navy's interest in tilt-duct V/STOL aircraft waned, and no X-22As ever entered service.

Deflected-Slipstream Concept

One of the earliest V/STOL concepts explored by Langley and Ames was the idea of using large wing flaps to deflect the slipstream of a propeller-driven aircraft downward to produce higher lift for takeoff and landing. If the propellers were made large enough to immerse the wing in their slipstream, it was reasoned that the beneficial power-induced effects would be quite large. The concept was attractive because it was not necessary to rotate wings or engines during the transition. Before such a design could be used for VTOL missions,

The Bell X-22 tilt-duct research aircraft in flight. The aircraft served the role of an in-flight variable-stability simulator for V/STOL aircraft designs and provided extensive information on handling-quality requirements and cockpit displays. (Photo courtesy of Smithsonian Institution via NASA)

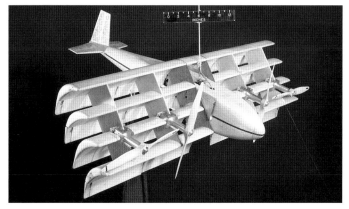

In the early 1950s Langley explored radical deflected slipstream configurations for vertical and/or short-takeoff and landing capability. This multi-engine deflected-slipstream model of 1952 used four wings and four propellers to induce lift by turning the propeller slipstream downward. Although lift was increased, large variations in trim, stability and control were produced, and free-flight tests of such configurations were judged to be unfeasible. (Photo courtesy of NASA)

some type of reaction control would have to be incorporated for low-speed control. The concept suffered from a basic limitation: the thrust produced by the engine was substantially reduced by the flow-turning process, whereas other V/STOL concepts—such as the tilt wing—do not experience any loss in thrust. Other serious limitations are that the turning of the slipstream by a wing trailing-edge flap induces a large nose-down diving moment on the aircraft; and that the effect of ground proximity can be very detrimental on induced lift. Nonetheless, the advent of high power-to-weight turboprop engines stimulated interest in the concept.

Langley initiated research on deflected-slipstream designs in the early 1950s with studies of "cascade-wing" aircraft having wing configurations that resembled deflectable venetian blinds. Conventional wind-tunnel force tests were made to measure aerodynamic performance, stability, and control. Geometric designs began to evolve, and researchers conducted a series of static (no wind) tests to determine the flow-turning efficiency of propellers in front of highly deflected multi-segment wing flaps. The resultant force on the model could be turned almost 90 degrees to a vertical orientation, while the thrust was reduced to about 80 percent of the propeller thrust. They also found that the presence of the ground reduced the turning angle to about 75 degrees. A large number of aerodynamic tests were conducted in the Langley Full-Scale Tunnel and the Langley 300-mph 7- by 10-Foot Tunnel, with emphasis on defining flow turning efficiency and optimum flap configurations.

A series of free-flight tests of various deflected-slipstream models ensued at Langley, and in the late 1950s the Ryan VZ-3 Vertiplane was developed and tested at NASA Ames as the first deflected-slipstream research aircraft. As expected, the vehicle suffered a lift deficiency in ground effect and could not make a vertical takeoff, but hovering flight was achieved out of ground effect at altitude following a short ground-run takeoff.

Fairchild VZ-5

The Fairchild VZ-5 was designed under an Army contract as a two-place, deflected-slipstream research aircraft in the 1950s. A single turboshaft engine drove four three-blade propellers with cross-shafting. Differential propeller pitch on the outboard propellers provided roll control in hovering flight, while tail fans provided yaw and pitch control. The aircraft never flew, although some tethered tests were made in the late 1950s.

Langley supported the VZ-5 program with tests in the 300-mph 7- by 10-foot tunnel and the full scale tunnel.

The Fairchild VZ-5 deflected-slipstream aircraft during tests in the full-scale tunnel in 1961. The configuration's aerodynamic characteristics were unacceptable in terms of stability, trim, and allowable range of center-of-gravity location. Note the tail-mounted fans used for control at low speeds. Flight testing was limited to a few tethered flights. (Photo courtesy of NASA)

The VZ-5 was tested in the full-scale tunnel in 1961, and the results of the test program were very disappointing. With the aircraft center of gravity in the design position, the aircraft was aerodynamically unstable over the speed range, and was incapable of being trimmed longitudinally at low speeds or in hovering flight conditions. In order to obtain stability, it would have been necessary to add an unacceptable ballast of about 700 pounds to the cockpit area. In addition, the aircraft exhibited very large values of effective dihedral, which would have likely resulted in unacceptable lateral-directional characteristics. As a result of these deficiencies, the VZ-5 never flew, and the project was terminated.

Ducted-Fan Concept

The Langley and Ames research centers have been hotbeds of V/STOL research on ducted-fan V/STOL concepts from the early 1950s through the current day. These configurations have ranged from single-pilot stand-on vehicles featuring kinesthetic control by the pilot, to sophisticated buried-fan, fan-in-wing, or fan-in-fuselage designs. The data generated by the NACA and NASA and their research partners during these exhaustive studies has been the fundamental foundation for the design and development of such vehicles. Even a brief review of all the significant research in this area is far beyond the authors' intentions, but an example of the research at Langley follows for the fan-in-wing Ryan XV-5A configuration.

Some of the most significant developments in V/STOL technology were the advanced lift-fan propulsion concepts by the General Electric Co. By combining tip-driven lift fans with a conventional turbojet, General Electric conceived a dual-propulsion concept for relatively high-speed V/STOL vehicles. The company developed large-scale lift fans through extensive wind-tunnel testing at Ames and other ground tests to the point that the U.S. Army Transportation Research Command was stimulated to fund the development and flight testing of two research aircraft in 1961.

Ryan XV-5A

With General Electric as the prime contractor and Ryan as the airframe partner, work on two research aircraft

The 0.18-scale free-flight model of the XV-5A fan-in-wing aircraft. For conventional wing-borne flight, fan cover doors were used to seal the wing-fan cavities, and vane segments sealed the nose-fan cavity. View shows the sophisticated tip-driven fans fabricated for the model. The model used a three-tube compressed-air ejector arrangement forward of each engine intake behind the cockpit to deliver air to the fans via internal ducting. The relatively sophisticated model was made of composite materials and flown in hovering and transitional flight. No attempt was made to investigate the conversion maneuver to the cruise configuration. (Photo courtesy of NASA)

The second XV-5A in conventional flight above Edwards Air Force Base. The first XV-5A was destroyed in a fatal crash that took Ryan test pilot Lou Everett's life during a transition maneuver in April 1965. The second aircraft was also involved in a fatal crash, but it was rebuilt and modified as the XV-5B and flown at NASA Ames for V/STOL research. (Photo courtesy of NASA)

designated XV-5A (originally VZ-11) began with support from NASA Langley. The XV-5A was a small fan-in-wing aircraft powered by two J85 turbojet engines located above and to the rear of the two-place cockpit. A pair of five-foot-diameter lift fans were buried in the wing panels, and a smaller three-foot diameter fan in the nose was used for pitch trim and control. For cruise flight, the jet engines' exhaust was routed through conventional tailpipes, and the fans were covered with auxiliary doors to form a smooth outer wing contour.

For conversion to hovering flight, the cover doors were opened and the pilot actuated a valve to divert the engine exhaust through a ducting system to drive the tip-driven wing fans, as well as the nose fan. The three fans were capable of producing about three times the total thrust of the two J85 engines that drove them. Located under each wing fan was a set of louvered vanes that could be deflected rearward to vector the fan thrust and thereby impart forward thrust for transition. When the vanes were deflected differentially between right and left fans, yaw control could be produced for hover. Finally, the vanes could be used to choke the flow from a lift fan, producing roll control. The nose fan included a "scoop door" that was deflected to control pitch. In high-speed flight, an all-movable horizontal tail in a T-tail arrangement, rudder, and ailerons provided control.

At Langley, wind-tunnel testing of powered XV-5A models was conducted in the full-scale tunnel and in the 300-mph 7- by 10-foot tunnel. In 1962 Langley fabricated a 0.18-scale free-flight model of the XV-5A at the Army's request and tested it in the full-scale tunnel. The model was built of composite materials and equipped with sophisticated tip-driven fans and an internal ducting system. The fans were tip-driven with compressed air, and the fan exhaust louvers could be deflected collectively for thrust-spoiling (altitude control) or differentially for roll and yaw control. However, for the free-flight investigation, jet-reaction controls at the wingtips and tail were used to simplify the piloting task. The nose scoop doors, however, were used for pitch trim. The tests were conducted for hovering flight and for transition speeds up to about 97 knots (full scale), at which speed the conversion to wing-borne flight was scheduled. No conversion maneuvers were attempted in the model flight tests.

As was the case for the propeller-driven V/STOL models previously tested, the XV-5A model exhibited unstable pitch and roll oscillations in hovering flight

with the controls fixed, but the motions were slow and easy to control. As transition speed increased, the model required more nose-down control from the nose fan, resulting in a lift loss of about 12 percent.

The first XV-5A aircraft flew on May 15, 1964, at Edwards Air Force Base, piloted by Ryan Chief Test Pilot Lou Everett, who the reader may remember was also the test pilot for the Flex-Wing and Fleep parawing vehicles discussed in Chapter 4. In April 1965, the first aircraft experienced a fatal crash apparently caused by an inadvertent pilot control input in a transition maneuver during an official demonstration. Both XV-5A aircraft had been demonstrating the low- and high-speed performance capabilities when Everett's aircraft suddenly nosed over and crashed. The second XV-5A later experienced a fatal crash in October 1966, when a pilot-operated rescue hoist was ingested into one of the wing fans, causing the aircraft to roll and begin descending. The pilot ejected, but he did so at an unsurvivable roll angle. The aircraft, which was not destroyed, was rebuilt with landing gear and cockpit modifications as the XV-5B. An extensive XV-5B flight-research program on aerodynamics, acoustics, and flying qualities was subsequently conducted by NASA at Ames.

Besides its very successful research efforts to help develop the vectored-thrust Harrier aircraft, Langley conducted model tests of other radical vectored-thrust V/STOL designs, such as this free-flight model of the proposed Republic-Fokker D-24 Alliance aircraft in 1961. The Alliance configuration was in response to competition for a NATO V/STOL supersonic strike fighter. The design featured variable-sweep wings and Harrier-like vectoring nozzles. The long, slender model was much too skittish and difficult to control in roll during hovering flight attempts in the Langley Full Scale Tunnel, and a planned flight-test program in the tunnel was cancelled. (Photo courtesy of NASA)

CHAPTER 6

CIVIL CONCERNS

Throughout its 90-year history, the Langley Research Center has focused research activity toward identifying and resolving the problems of flight for both military and civil aviation. As discussed in the introduction of this book, the center's role in serving the civil aeronautical community has significantly differed from its actions within the military community. Langley provides the civil aviation industry and other government agencies such as the Federal Aviation Administration (FAA) and National Transportation Safety Board (NTSB) with critical data and tools for the design and analysis of advanced aircraft configurations; analysis of the environmental impact of aircraft operations; concepts for enhanced safety and significantly improved performance; and access to NASA facilities (on a fee-paying basis) for developmental work during aircraft programs. This relationship with the civil industry has been especially beneficial for the general public and has helped maintain the American dominance of the commercial and general-aviation aircraft markets. Arguably, an even more important contribution of the agency has been in contributing technology and expertise in efforts to improve the safety and productivity of this nation's public air-transportation system.

Discussion of Langley's numerous successful contributions for the civil community is not the primary intent of this book; however, a considerable amount of information is available to the reader via the literature, NASA publications, and web-based information sites. Information is available, for example, on Langley's contributions to aviation safety, which include conception of the wide-spread use of runway grooving, in-flight wind-shear detection and avoidance, non-intrusive structural inspections, flutter avoidance, VGH (velocity, load factor, and altitude) instrumentation, prediction of gust loads, advanced "glass" cockpit displays, wake-vortex hazard characterization and avoidance, stall/spin technology, and aircraft crashworthiness concepts. Within the area of environmental concerns, researchers have contributed leading-edge data and design methods for aircraft and propulsion-system noise prediction and reduction, as well as the impact of aircraft operations on the atmosphere. Langley's contributions to enhanced aircraft performance have included leadership in the development of weight-saving composite materials, winglets, low-speed and supercritical airfoils, the NACA cowling, high-lift devices, drag-reduction techniques, and radical aircraft configurations for civil applications.

Finally, the research center has provided the U.S. industry with powerful design tools, including advanced computer software for design analysis and unique wind tunnels and other test facilities. Many of the test facilities are made available to the civil aircraft industry under fee-paying arrangements for development support during emerging aircraft programs.

The purpose of this chapter is to discuss major research efforts at Langley on radical civil aircraft that have not yet reached production status. These activities have involved large efforts on very difficult technical problems that have been alleviated by many degrees; however, the concepts have not advanced because of remaining obstacles such as excessive cost, lack of compliance with current and projected environmental regulations, or lack of market demands.

The Sonic Transport

After the introduction of commercial jet transports during the 1950s, the quest for improved aircraft performance followed the usual "higher, faster, farther" goals within the design community. With the advent of swept wings and jet propulsion, tremendous strides were made in operational envelopes for transports, and researchers and designers continued to push for the next major breakthrough in technology.

The introduction of the Boeing 707 and other early jet transports provided cruise speeds on the order of Mach 0.7 to Mach 0.8. When Langley's Dr. Richard T. Whitcomb conceived and developed the supercritical airfoil with its performance-enhancing characteristics in the 1970s, it appeared that an unprecedented opportunity might appear for transports that could cruise at substantially higher speeds, even approaching Mach 1. The drag-reducing aerodynamic performance improvements provided by Whitcomb's new supercritical airfoil technology significantly delayed the onset of the classical drag-rise Mach number, delayed the onset of high-speed buffet that results from wing-flow separation induced by the formation of shock waves, and also improved high-lift performance at the low speeds used for approach and landing.

NASA initiated an Advanced Technology Transport (ATT) Program in mid-1970 to explore the feasibility and technology required for a subsonic, long-haul civil transport that could cruise near the speed of sound. Whitcomb led an aggressive program to explore aerody-

namic design guidelines for this radical new class of transports. In addition to his supercritical wing concept, he had previously conceived the brilliant "area-rule" concept in the 1950s, which provided theory and design approaches for a dramatic reduction in drag for aircraft in the transonic speed range. Armed with these two powerful concepts, Whitcomb and his staff explored the aerodynamic performance characteristics of advanced new transport configurations that combined supercritical wings and fuselage/wing area ruling. Several generic models were tested in the Langley 8-Foot Transonic Pressure Tunnel, and encouraging tunnel test

Langley's Richard T. Whitcomb and a model of a near-sonic transport configuration in 1970. Whitcomb's careful sculpting of aircraft shapes for improved performance at high-subsonic speeds is legendary. Note the area-ruled wing root and fuselage lines. (Photo courtesy of NASA)

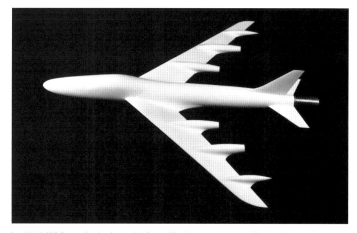

In 1959 Whitcomb designed this radical transport configuration, which used fuselage contouring and shock bodies on the wing to delay the onset of transonic drag. (Photo courtesy of NASA)

results indicated the potential for near-sonic cruise capability. The carefully tailored aircraft configurations were sleek, futuristic visions that were considerably different from conventional jet transports. In particular, the integration of area ruling for transport aircraft resulted in markedly different configurations.

NASA's ATT Program awarded contracts to Boeing, McDonnell Douglas, and General Dynamics for a preliminary analysis of the advantages and disadvantages of representative near-sonic transports. A key input for the studies was aerodynamic data for a representative sonic transport configuration obtained by Whitcomb's staff in the 8-Foot Transonic Pressure Tunnel. The industry studies' results indicated that a sonic transport having aerodynamic characteristics similar to the Langley data would have an increase in cruise speed of about 20 percent and be capable of cruising at a Mach number of about 0.98 with operating costs similar to those of conventional subsonic jet transports. Boeing was so impressed by the potential for new configuration that it actually included a candidate design during its preliminary design of the 767 transport.

Unfortunately, advances in technology are frequently restrained by non-technical political and economic issues. About two months after the industry studies were completed, the Organization of Petroleum Exporting Countries (OPEC) tripled the price of crude oil. The airline community's response of the airline community was an immediate decrease in interest in flying faster. Instead, fuel consumption became the key parameter for aircraft sales, and the commercial aircraft industry recommended that NASA decrease its sonic transport effort and redirect activities toward improved efficiency at lower cruise speeds. NASA renamed its program the Advanced Transport Technology Program, and Whitcomb and his team redirected their efforts to obtain more aerodynamic efficiency through use of supercritical technology. As design-trade studies continued within the industry, an approach emerged in which the benefits provided by supercritical wing technology were used to increase the thickness of the wing, reduce the wing sweep to reduce structural weight, and increase the aspect ratio of the wing to reduce drag. Using this approach, the commercial jet transport industry has consistently produced products with unprecedented fuel efficiency.

It is important to note that the application of Whitcomb's supercritical wing technology ultimately involved a different use of the approach than was intended by Whitcomb when he conceived the concept. The fact that an idea to promote flight at near-sonic speeds was ultimately used on a world-wide basis to allow transports to cruise more efficiently at lower speeds is not untypical of research projects. This experience is a vivid demonstration of the reality that it is often impossible to identify all the real-world applications and justifications in advance for a basic research activity.

Boeing briefly revisited the concept of a near-sonic cruise transport in 2001 when it announced the existence of its "Sonic Cruiser" design project. The aircraft was designed to cruise at Mach 0.95 with a 20-percent faster cruise speed than conventional contemporary transports. However, it would also burn fuel at a higher rate, and the airlines once again opted for more fuel-efficient designs that would cruise at lower Mach numbers. The project was terminated in 2002 without any airline orders for production.

Supersonic Civil Aircraft

The introduction of the Boeing 707 in the 1950s resulted in a revolution in the transport industry and in the public's expectations for air travel. The inherent value of increased speed and reductions in travel time (especially for business travelers) sparked an embryonic interest in the possibility of supersonic civil transports.

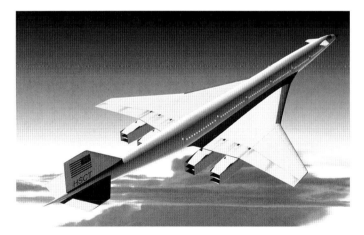

An economically and environmentally viable supersonic transport has been the goal of numerous domestic and international efforts since the 1960s. Despite extensive research efforts, the dream has been continually extinguished by constraints posed by unacceptable levels of sonic booms, takeoff noise, and rising fuel prices. This artist's rendering shows a hypothetical high-speed civil transport (HSCT) in flight. (Graphic courtesy of NASA)

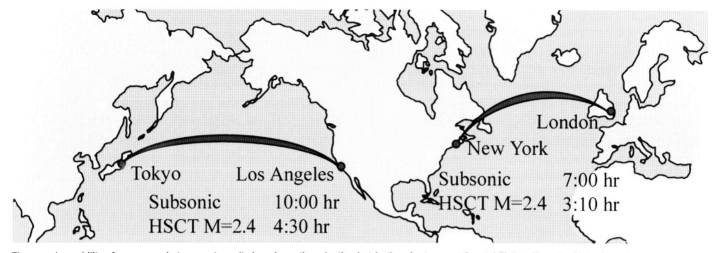

The speed capability of a supersonic transport results in a dramatic reduction in trip time for transcontinental flights. Here the flight times for representative trips are compared for a conventional subsonic transport and a high-speed civil transport with a cruise Mach number of 2.4. (Graphic courtesy of NASA)

Visionary designers began to advance paper studies of supersonic transports that could fly 300 passengers at more than 1,600 mph (Mach 2.4), speeding across the Pacific from Los Angeles to Tokyo in 4.5 hours. Unfortunately, the very large amounts of monetary and human resources expended in the United States and abroad to advance the technologies required for financially viable and environmentally compatible supersonic civil transport operations have failed.

The pursuit of a successful supersonic civil transport has consumed more aeronautics resources within NASA than any other aeronautics research activity. These activities have included extensive participation by all the NASA aeronautics centers, industry, the military, and universities for more than 45 years and have consisted of studies ranging from fundamental flow physics to propulsion systems and flight tests of radical aircraft. At Langley, the supersonic transport effort has involved five different eras of interest.

Langley's first efforts in supersonic research began during its days as a NACA laboratory in the late 1930s. Researchers focused on studies of the flow physics associated with supersonic flight and the supersonic aerodynamics of generic shapes and aircraft configurations. Langley evolved a substantial supersonic aerodynamic database and conceived, constructed, and maintained new transonic and supersonic wind tunnels and other facilities. The well-known contributions of R. T. Jones on the benefits of wing sweep, participation in the breakthrough X-1 program and other high-speed X-plane programs, Whitcomb's area rule, the development of the slotted wind tunnel which permitted accurate tests

at sonic conditions, and a vast amount of general supersonic wind-tunnel and flight studies placed Langley researchers at the forefront of supersonic technology. Building on its substantial scientific foundation, NACA participated with the DoD in the development and subsequent demonstration of supersonic flight with research aircraft and the famous "Century Series" fighters in the late 1950s. In addition, Langley researchers conducted pioneering research on a new phenomenon that would prove to be a major barrier to the development of successful supersonic civil transports: the creation and propagation of sonic booms.

The U.S. Supersonic Transport Program

The second era of Langley's supersonic research began with the formation of NASA in 1958 and lasted until 1971. Two projects dominated these research activities: the supersonic cruise XB-70 bomber and the U.S. Supersonic Transport Program. The XB-70 was the most important early activity to stimulate supersonic transport research at Langley. The XB-70 program had begun in 1957 and was canceled a few years later, but two XB-70 aircraft were completed for research flights. In view of the similarity of its relative size and cruise speed to a representative supersonic civil transport, the XB-70 evoked considerable interest within NASA for research relevant to civil applications. In addition to supporting the military's XB-70 developmental efforts with traditional high-priority tests in virtually all of its wind tunnels prior to the cancellation of the program, Langley began advocating for a research program for civil transports in the mid-1960s. After the crash of the second XB-70 in a midair collision at Edwards Air Force Base on June 8, 1966, NASA conducted aerodynamic, propulsion, handling-quality, and sonic-boom research with the first aircraft.

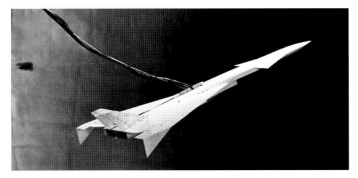

Langley's role in support of the XB-70 program stimulated interest in civil supersonic transports. Extensive wind-tunnel, simulator, and flight tests were conducted at the NACA and NASA centers. This photograph shows free-flight tests of an early version of the XB-70 configuration in the Langley Full Scale Tunnel in 1957 in support of what was then known as Air Force Weapon System WS-110A. The familiar elegant lines of the configuration and its down-turned wingtips are apparent. (Photo courtesy of NASA)

Takeoff of the first XB-70 prototype with a TB-58 chase aircraft at Edwards Air Force Base in the 1960s. The second prototype had been selected for use in a national study of sonic-boom phenomena with NASA as a major participant. However, after the second XB-70 was destroyed in a mid-air collision, NASA was assigned management of the first vehicle, which was then used for studies of stability and control, turbulence response, sonic-boom characteristics, aerodynamic performance, and takeoff and landing noise. The XB-70 flight program began in September 1964 and lasted until February 1969. The aircraft is now on display at the National Museum of the U.S. Air Force at Wright-Patterson Air Force Base. (Photo courtesy of NASA)

Meanwhile, NASA Langley leaders held numerous discussions with the FAA and the DoD to formulate a cooperative American SST Program and to define agency responsibilities within such a program. President Kennedy subsequently assigned leadership of the program to the FAA, with NASA providing basic research and technical support. With its cadre of leading experts in supersonic aerodynamics, Langley was poised to propose promising configurations for supersonic transports to the national team. Configuration studies within NASA led to wind-tunnel tests beginning in 1959 of 19 different NASA-conceived SST designs referred to as supersonic commercial air transport (SCAT) configurations. Testing continued for more than seven years on 40 variants of these designs at Langley in its subsonic, transonic, and supersonic wind tunnels.

In 1963, NASA awarded contracts to Boeing and Lockheed to conduct industry-feasibility studies of four potential NASA candidate configurations known as SCAT-4, SCAT-15, SCAT-16 and SCAT-17. The SCAT-4 was an elegant arrow-wing design, SCAT-15 was an innovative variable-sweep arrow-wing design that used auxiliary variable-sweep wing panels, SCAT-16 was a more conventional variable-sweep design, and SCAT-17 was a delta-canard design. The Lockheed and Boeing

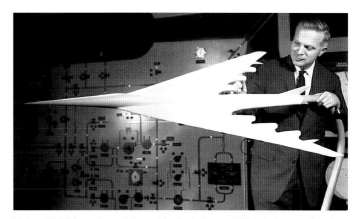

Richard T. Whitcomb participated in the in-house NASA design studies for candidate supersonic transport configurations. Shown here with one of his designs, he directed the design and wind-tunnel testing of a NASA candidate known as the SCAT-4. Later, when industry assessed the economic feasibility of the family of NASA designs, Whitcomb was discouraged by his pessimistic personal outlook for such an aircraft's success. He subsequently left the SST-focused efforts at Langley in disgust and redirected his efforts to advanced subsonic transports, where he then conceived the highly successful supercritical airfoil and winglet concepts. In his later years, Whitcomb was still outspoken against the feasibility of supersonic transports. (Photo courtesy of NASA)

studies concluded that the SCAT-16 and SCAT-17 designs were the most promising configurations.

Although the results of the 1963 industry studies were encouraging, it was generally conceded by all participants that the technology for a viable supersonic transport was not yet available. Nonetheless, the activity occurred at a time when the United States was reeling from an aura of technical inferiority brought about by the Soviet Union's launch of *Sputnik* and the European announcement of a startup for a new civil commercial supersonic transport to be known as Concorde. In this atmosphere, on June 5, 1963, President Kennedy delivered a speech before the graduating class of the U.S. Air Force Academy in which he committed the nation to "develop at the earliest practical date the prototype of a commercially successful supersonic transport superior to that being built in any other country in the world..."

The FAA issued a request for proposals to industry for a U.S. supersonic transport having a cruise Mach number of 2.7, a titanium structure, and a payload of 250 passengers. Boeing chose a variable-sweep wing configuration as its entry in the competition (Model 733-197), and Lockheed chose a fixed-wing, double-delta design (Model CL-823), on the basis that it would be a simpler, lighter airplane. As weight problems began to appear for both designs during the design cycle, the FAA advised both Boeing and Lockheed to explore the NASA SCAT-15 design as a potential alternate.

Extensive research activities in support of the SST program took place at Langley, including wind-tunnel tests, configuration assessments, and disciplinary studies and consultation in aerodynamics, structures, materials, flight controls, noise, sonic boom, propulsion integration, aeroelasticity, landing loads, stability and control, operating problems, integration of SST and subsonic air traffic, and flight dynamics (including free-flight models). In addition, Langley researchers initiated the development of advanced computational tools and theory for the design and evaluation of complex supersonic configurations. These contributions included the first application of high-speed computers to the aerodynamic design of supersonic aircraft. Using such tools, designers were able to shorten the time

The original SCAT-15 configuration included auxiliary variable-sweep wing panels for improved low-speed aerodynamic characteristics. Industry analyses indicated that the configuration would have excessive weight and other issues. Although directed by the FAA to study the configuration as an alternative, Boeing did not fully endorse the design. (Photo courtesy of NASA)

Desk-top models of the NASA SCAT configurations. Clockwise from the lower left are the radical variable-sweep SCAT-15, the more conventional variable-sweep SCAT16, the canard-delta wing SCAT-17, and the streamlined SCAT-4. Langley researchers designed all of the configurations, with the exception of the SCAT-17, which was an Ames design. (Photo courtesy of NASA)

The Langley Full Scale Tunnel was the scene of numerous low-speed studies of supersonic transport configurations from the 1960s through the 1990s. This photograph shows a free-flight test of the Boeing variable-sweep SST model in 1967. (Photo courtesy of NASA)

required for the prediction of critical aerodynamic parameters such as supersonic wave drag from more than three months to just a few days (for a single Mach number and for a single configuration).

After several design iterations, Boeing was ultimately declared the competition winner on December 31, 1966. The company then changed its design to a fixed-wing, double-delta SST configuration. Boeing did, however, continue its studies of the SCAT-15 design. In its role as primary advocate for the SCAT-15 design, Langley pursued improved versions of its own original SCAT concept as a potential alternative to the Boeing double-delta design. In 1964, researchers used their new computational tools to design an improved derivative, called the SCAT-15F. This computer-generated, fixed-wing version of the earlier variable-sweep SCAT-15 configuration was demonstrated by wind-tunnel tests to exhibit a lift-to-drag ratio of 9.3 at Mach 2.6, an impressive 25- to 30-percent improvement over the state of the art at that time. In 1966, the FAA requested that NASA conduct more in-depth studies of the SCAT-15F design, and the project quickly became a high-priority activity at Langley, with extensive computational analysis and wind-tunnel testing of several models in virtually every subsonic, transonic, and supersonic tunnel at Langley.

As in many past studies of radical configurations, the Langley Full Scale Tunnel was a major workhorse for the development of the SCAT-15F configuration. Free-flight models were once again used to evaluate dynamic stability and control, augmented by piloted ground-based simulator studies of the handling qualities of the SCAT-15F during approach and landing. A major problem identified for the SCAT-15F design involved deficient low-speed stability and control characteristics. The highly swept, arrow-wing configuration exhibited a longitudinal instability (pitch-up) tendency typically shown by arrow wings at moderate angles of attack, and the instability was accompanied by the possibility of a dangerous unrecoverable "deep stall" behavior. The deep stall for the SCAT-15F occurred for angles of attack slightly higher than those for the pitch-up tendency, and was characterized by an abrupt increase in airplane angle-of-attack to extremely high values (on the order of 60 degrees), where longitudinal controls were ineffective for recovery. Several Langley tunnel entries were directed at the unacceptable pitch-up problem and the development of modifications to alleviate it. After

extensive testing in the full-scale tunnel, a combination of modifications—including a 60-degree deflection of the wing leading-edge flap segments forward of the center of gravity, a "notched" wing apex, Fowler flaps, and a small aft horizontal tail—eliminated the deep-stall trim problem. In today's world, instabilities like that exhibited by the SCAT-15F are routinely eliminated by automatic control systems that prevent the pilot from

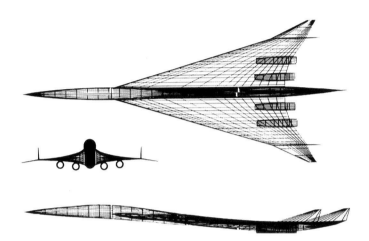

One of the most valuable Langley contributions in the U.S. SST program was the first application of high-speed computers to the design of supersonic aircraft. The use of these computer tools shortened analysis time from months to days. The graphic shows a computer representation of the SCAT-15F variant of the SCAT-15 design. (Photo courtesy of NASA)

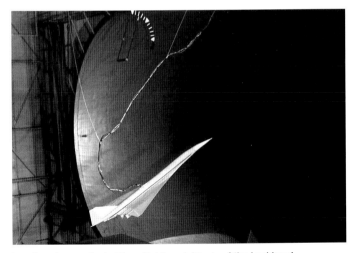

Langley also conducted free-flight model tests of the Lockheed double-delta SST configuration in 1967. Aerodynamic characteristics of the Boeing and Lockheed models were also measured in conventional wind-tunnel tests at the facility to obtain aerodynamic data for analysis of the free-flight motions as well as to serve as inputs for Langley-piloted simulator studies of the full-scale aircraft's handling qualities. (Photo courtesy of NASA)

accidentally entering the condition, but the technology was not ready for civil applications in the early 1960s.

Arguably, the most critical technical barrier for an American SST (and subsequent supersonic transport programs) was the concern over potential impacts of the

A wind-tunnel technician prepares a model of the SCAT-15F for supersonic flow visualization tests in the supersonic Langley Unitary Plan Wind Tunnel in 1970. The SCAT-15F exhibited outstanding supersonic aerodynamic performance and remains one of the most efficient supersonic-cruise configurations to this day. Low-speed stability and control issues discouraged industry acceptance of the design as an alternate SST option. (Photo courtesy of NASA)

The full-scale tunnel was a major workhorse during refinement studies for the SCAT-15F configuration. Shown here, the free-flight model of the supersonic transport design is mounted to a wind-tunnel support system used to vary the angle of attack and angle of sideslip during tests. Note the electrical wires taped to the support sting entering the rear of the model. These leads transmit data signals from an electrical strain gauge balance in the model. Several intensive tunnel entries were made in efforts to provide solutions to a longitudinal pitch-up problem exhibited by the design at low speeds representative of takeoff and landing. Numerous geometric changes were made to the model, and after extensive testing an acceptable solution was identified. (Photo courtesy of NASA)

sonic boom. Langley researchers in noise technology were invited to participate in joint NASA-DoD assessments to determine the level of sonic-boom exposures that might be accepted by the public. The group at conducted sonic-boom flight tests St. Louis from November 1961 through January 1962, at Edwards Air Force Base in 1963, at Oklahoma City in 1964, and at Chicago in 1965. In 1966 and 1967, overland flights of XB-70, F-104, F-106, B-58, and SR-71 aircraft at Edwards Air Force Base were conducted to study human acceptance of sonic booms and to measure the response of typical house structures to various levels of overpressure under the sponsorship of The Office of Science and Technology. More than 350 supersonic overland flights were conducted, measurements of boom overpressures were made, and subjective human assessments of the acceptability of the booms were obtained. The results of the flight tests clearly indicated that unexpected, abrupt levels of overpressure as low as 1.5 to 2.0 pounds per square foot were unacceptable to human subjects. All these data and additional surveys led the aviation industry and the FAA to conclude that operations of a supersonic civil transport would probably be constrained to supersonic flight over water only, or supersonic over land in low-population corridors.

Faced with extensive domestic controversy, technical issues, international politics, and a growing public outcry over sonic booms, airport noise levels, and other environmental concerns, the U.S. Congress canceled the SST program in March 1971. After eight years of research and development and an expenditure of approximately $1 billion, the United States withdrew from the international supersonic transport competition. Today, many participants who were close to the program consider the cancellation a wise decision because of the immaturity of technology, bleak profitability outlook, and high risk of failure.

The NASA Supersonic Cruise Research (SCR) Program

The third era of Langley's research occurred between 1971 and 1981. Following the termination of the American SST Program by Congress, NASA initiated a new area of study known as the NASA Supersonic Cruise Research (SCR) Program to continue research on the problems and technology of supersonic flight. The new program's goals were to build on the knowledge gained during the SST program and to provide the supersonic

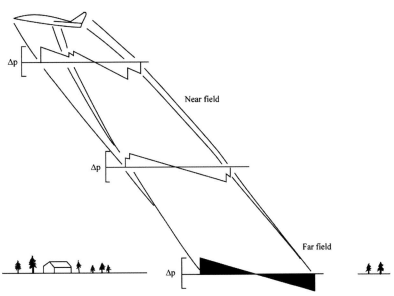

Shock waves emanate from an aircraft's various components in supersonic flight. Near the aircraft, the pressure differences through the shock waves are notably different and distinguishable; however, at ground level in the far field the pressures tend to coalesce into a single abrupt pressure rise followed by an abrupt pressure decrease (called an "N" wave) recognizable as a sonic boom. If the differential pressures are of sufficient magnitude, structural damage may occur to buildings or other objects in the flight path. Even without damage, the "startle factor" of the boom noise may be unacceptable to humans or animals. Today, the law prohibits supersonic flights by civil aircraft over the United States. (Graphic courtesy of NASA)

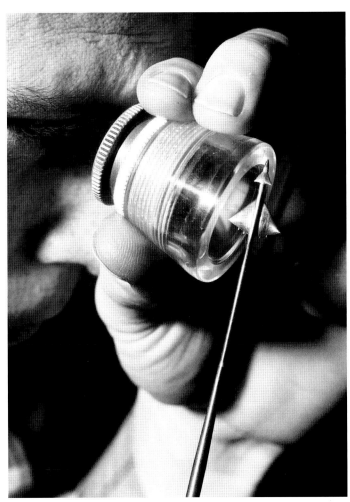

During the U.S. supersonic-transport program, Langley researchers developed a wind-tunnel testing technique in the supersonic unitary plan tunnel to measure far-field pressure differentials caused by shock waves. Due to the relatively small size of the supersonic tunnel test section, very small models were required for representation of the far-field pressure characteristics. In this photo, a researcher is inspecting a representative model used to generate the flow phenomenon. (Photo courtesy of NASA)

technology base that would permit the United States to keep its options open for proceeding with the development of an advanced supersonic transport, if and when it was determined to be in the national interest. However, the SCR Program's pace (limited to four years) and funding were intentionally limited by Congress to avoid another SST battle. The SCR activity involved all four of the NASA aeronautical research centers and many aerospace companies, research organizations, and universities. Although NASA Headquarters managed the program, Langley led the day-to-day operations with research tasks being accomplished at NASA centers and by industry. The program's disciplinary research included aerodynamic-performance studies at Langley and Ames, propulsion studies at Lewis (now Glenn Research Center, Ohio), structures and materials studies at Langley, Dryden, and Ames, stability and control studies at Ames, Dryden, and Langley, and studies of stratospheric emissions impact at all centers.

By the time the U.S. SST program ended in 1971, supersonic aerodynamic design methods were well in hand, as evidenced by the remarkable aerodynamic efficiency of the SCAT-15F design. However, efficient, highly swept supersonic configurations still exhibited deficiencies in subsonic stability, control, and performance. Thus, a major effort was directed to improve the low-speed behavior of supersonic transports during the SCR Program, and only a relatively limited effort was expended on improving supersonic cruise efficiency. In the full-scale tunnel, researchers explored wing planform effects, various wing leading-edge devices, and other innovative technical concepts in efforts to augment lift while providing satisfactory stability and control. Collaborative efforts with industry teams led to solutions for many of these problems.

Although the SCR Program could not afford expensive full-scale aircraft flight testing, researchers at Langley continued to refine wind-tunnel and computational methods to predict and minimize the effects of sonic booms. As progress continued, the concept of using radical aircraft shaping methods to modify the objectionable characteristics of sonic-boom signatures was pursued. Prior to the Langley studies, the scientific community had conducted studies that indicated the critical characteristics of the sonic boom signature, such as rise time and waveform of the overpressures, could be modified to result in more acceptable perceptions of loudness. Unfortunately, the SCR Program was terminated just as encouraging new Langley results were forthcoming from theory and wind-tunnel tests demonstrating that, at least in the near term, an aircraft's geometric features might alter pressure fields. NASA funding for sonic-boom research was then dropped and remained dormant for nearly six years.

Researchers also examined radical versions of other supersonic transport concepts during the SCR program. One such idea was the use of a twin-fuselage configuration for potential high-payload, high-productivity applications. This approach used dual fuselages and an unswept, long-chord wing section between the fuselages to increase internal volume. The aerodynamic principle promoted for the twin-fuselage supersonic transport was that the fuselage forebody compression pressures ema-

nating from each forebody would impinge on the rearward-facing surface of the half portion of the adjacent body, thereby increasing the afterbody pressure and reducing the drag by as much as 25- to 30-percent more than a single-fuselage configuration. In addition, the SCR program stimulated an interest in smaller supersonic business jets that might be able to minimize the sonic boom issue (boom strength is proportional to vehicle size). Although no production vehicles came from the business-jet studies, the subject would be revisited in later years.

Other radical supersonic transport configurations were explored in the SCR efforts. This artist interpretation of a twin-fuselage design is representative of high-capacity configurations studied within Langley's basic research program. Promising favorable supersonic aerodynamic interactions and a relatively large internal volume, such configurations were proposed and studied for merit. (Graphic courtesy of NASA)

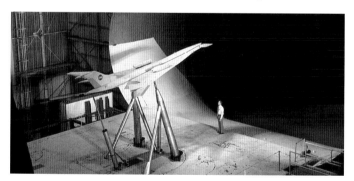

Far more wind-tunnel hours were expended in the NASA Supersonic Cruise Research Program in efforts to solve low-speed stability and control and performance issues than were used for supersonic aerodynamic studies. This photograph shows a large model of a supersonic transport configuration during tests in the Langley Full Scale Tunnel to determine the effects of upper-surface mounted engines on low-speed lift and stability characteristics. Although benefits were identified for lift augmentation (particularly with vectoring nozzles on the engines) and noise shielding during takeoffs and approach to landings, the supersonic aerodynamic interactions of the configuration were unfavorable. (Photo courtesy of NASA)

As industry grappled with the sonic-boom barrier, it was known that a smaller supersonic aircraft, such as a business jet, would have a lower boom overpressure level and might be more acceptable for over-land flights. (Photo courtesy of NASA)

As the 1970s drew to a close, the timing was wrong for any advocacy for supersonic transports. Anti-SST feelings still ran high in environmental groups, fuel prices soared, the Concorde was an economic disaster, and low-fare availability on wide-body subsonic transports was the public rage, rather than exorbitantly expensive tickets for supersonic junkets. In view of the situation and mounting funding problems for its Space Shuttle program, NASA terminated the SCR Program in 1982.

The data, concepts, and design methodology derived from aerodynamic research at Langley in the SCR Program proved invaluable in other NASA and DoD programs. One outstanding example of the value of Langley's supersonic aerodynamic-design expertise and methods was the cooperative venture with General Dynamics (now Lockheed Martin) for a joint design effort to develop a supersonic cruise wing for the F-16 fighter. As will be discussed in a later chapter, this joint activity produced the highly successful F-16XL version of the fighter.

The NASA High-Speed Research (HSR) Program

The fourth era of supersonic research at Langley began in the early 1990s with a program known as the NASA High-Speed Research (HSR) Program, which Langley managed for the agency. In the mid-1980s, interest in military and civil applications of hypersonic technology emerged as a new frontier in aeronautics, stoked along by President Ronald Reagan's 1986 State of the Union address in which he relayed visions of a Mach 25 hypersonic civil transport known as the "Orient Express." In response to this impetus, NASA awarded contracts to industry for studies of the market potential of a high-speed civil transport (HSCT) and the agency also conducted in-house studies to identify the most rational cruise speed for high-speed flight based on technology readiness and infrastructure costs.

The results of the industry and NASA studies concluded that a Mach 25 transport was not economically feasible, but a supersonic transport with a cruise Mach number between 2.0 and 3.2 was technically realizable if the well-known environmental issues of noise, sonic boom, and emissions could be solved. With congressional support, NASA initiated the NASA High-Speed Research (HSR) Program in 1990 to identify and develop solutions to the many environmental concerns sur-

rounding a second-generation supersonic transport. Initially the research focus was on the most important barriers to acceptance of supersonic transports, including issues on depletion of the Earth's ozone layers, excessive airport and community noise, and the unacceptable sonic boom. The hypothetical HSCT envisioned would carry 300 passengers at a cruise speed of Mach 2.4 and could cross the Pacific or Atlantic in less

Langley researcher David E. Hahne poses with a 0.10-scale model of a conceptual supersonic transport configuration during tests in the Langley Full Scale Tunnel in 1992. This configuration was one of several considered in the NASA High-Speed Research Program having the capability to fly 300 passengers at Mach 2.4 across the Pacific from the U.S. West Coast to Tokyo in about four hours. As was the case for earlier SST programs, off-design aerodynamic characteristics significantly impacted landing, takeoff, and subsonic performance characteristics. Several concepts were explored in this test program to improve capabilities, including a canard and a wing leading-edge flap system. (Photo courtesy of NASA)

than half the time presently required on modern subsonic, wide-bodied jets – at an affordable ticket price (estimated at less than 20 percent more than comparable subsonic flights), and be environmentally friendly.

After the first phase of HSR activities, sufficient information had been advanced to indicate that solutions to many of the problems were eminent, but the sonic-boom issue could not be resolved, and so the vehicle would be limited to supersonic flight over water only. The next phase of research attacked the issues of technology readiness and economic viability. NASA's industry partners were the Boeing Commercial Airplane Group, McDonnell Douglas Aerospace, Rockwell North American Aircraft Division, General Electric Aircraft Engines, and Pratt & Whitney. During this phase of the HSR program, NASA was spending one-quarter of its annual $1 billion aeronautics budget for high-speed research.

In order to focus the research efforts, the HSR participants agreed to a baseline Mach 2.4 vehicle concept known as Reference H designed by Boeing. Extensive wind-tunnel testing of the configuration was conducted in all of Langley's tunnels including subsonic, transonic, and supersonic facilities. By late 1995, results in all technical discipline areas continued to advance the state of the art, and a new updated baseline known as the Technology-Concept Airplane (TCA) was adopted.

Under the sponsorship of the HSR program, renewed interest and funding support began to accelerate as NASA researchers, industry, other government agencies, and academia began to examine theoretical aspects of sonic-boom propagation, experimental results of boom propagation, results of boom-acceptability studies for humans and animals (including sea life), and low-boom configuration design, analysis, and testing. Langley's researchers returned to the sonic-boom efforts they had terminated when the SCR program had ended. They quickly revisited the potential benefits of modifying the sonic-boom signature by altering the geometry of the generating aircraft. These complex studies addressed shockwaves generated by aircraft components such as the fuselage, wing, nacelles, and tails. During 1990, wing-body-nacelle models were used in the unitary plan tunnel to demonstrate the capability of modifying signatures. Researchers were also able to use emerging computational fluid dynamics (CFD) methods for guidance in developing configuration features to help tailor the sonic-boom signature.

Research activities on sonic booms within the HSR program were, of course, coordinated among NASA aeronautics research centers. In particular, the Langley Research Center and the Dryden Flight Research Center participated in joint assessments of technology development in the

As the NASA High-Speed Research Program progressed, the industry and NASA participants narrowed their configurations focus down to a design by Boeing known as Reference H. In this photograph from 1993, a model of the Mach 2.4 Reference H transport is being prepared for takeoff- and landing-performance testing in the Langley 14- by 22-Foot Subsonic Tunnel. (Photo courtesy of NASA)

A 0.015-scale model of the Technology Concept Airplane (TCA) configuration is tested in the Langley 16-Foot Transonic Tunnel for research on propulsion integration in 1999. A major design challenge for supersonic transport configurations is the efficient use of engine nacelles and noise suppressors while maintaining aerodynamic performance across the speed range. (Photo courtesy of NASA)

wind tunnels and in flight. During 1995, a NASA team at Dryden conducted an extremely successful series of flights with a NASA SR-71 to study the characteristics of sonic booms. Two other NASA research aircraft were used to obtain in-flight sonic-boom measurements. A NASA YO-3A propeller-driven "quiet" airplane flew subsonically between the SR-71 and the ground, while a NASA F-16XL flew supersonically near the SR-71 to measure the sonic boom at distances as close as 85 feet below and to the rear of the generating aircraft. Speeds flown by the two aircraft ranged between Mach 1.25 and Mach 1.8 at altitudes of about 30,000 feet.

In the late 1990s, Boeing's interest in continuing its level of effort in the HSR program dramatically diminished because of high-risk technical issues, subsonic commercial transport business demands, and marketability issues. The company subsequently announced its withdrawal from the HSR effort, stating its perspective that supersonic transports could not reasonably be ready for the marketplace before the year 2020. In response to the loss of its major partner in supersonic transport research, NASA terminated the HSR program in February 1999.

Back to Basics

The year 1999 marked the beginning of a series of events that severely dampened interest in supersonic-transport technology. First, the NASA HSR program was canceled. This was followed by the retirement of the

In 1995 the NASA Dryden Flight Research Center conducted an innovative flight-test study to measure the sonic-boom characteristics of an SR-71 generator aircraft at supersonic speeds. Shown here with the SR-71 is a NASA F-16XL research aircraft used to measure the boom properties of the SR-71 in the near field at altitudes of about 30,000 feet. (Photo courtesy of NASA)

Concorde fleet in 2003, then fuel prices soared to unprecedented levels, and many commercial airlines fell by the wayside. Langley's research activities were severely reduced in scope and funding, and the remaining funds and scientists were reoriented toward more fundamental research on a few critical technologies. Given the constraints of the day, researchers once again focused their precious resources on the Holy Grail of supersonic flight: minimization of the sonic-boom problem.

Alteration of the sonic-boom signature by use of aircraft geometrical shaping techniques was proposed almost 50 years ago, but had not been demonstrated in real-world conditions with an aircraft in supersonic flight. Researchers theorized that if the typical sharp-edged "N" waveform associated with the sonic-boom overpressures in the far field could be modified so that the onset and peak level of the initial shock was reduced and shaped (lower rise time or less abrupt), the booms might be less objectionable to human beings. Langley researchers had verified the benefits of signature shaping by using human test subjects exposed to simulated shaped sonic booms in a high-fidelity acoustic facility.

It is well known that the magnitude of sonic boom overpressure is related to primary aircraft features such as size, weight, and length. In addition, smaller aircraft exhibit a lower level of sonic boom than large transports. Coupled with promising research on boom-signature shaping via aircraft geometric configurations, these facts stimulated interest in the possibility of small, supersonic, business-category jets. In 2000, the Defense Advanced Research Projects Agency (DARPA) initiated a Quiet Supersonic Platform (QSP) Program directed toward the development and validation of critical technology for long-range advanced supersonic aircraft with substantially reduced sonic booms, reduced takeoff and landing noise, and increased efficiency relative to current-technology supersonic aircraft. The design's quantitative requirements appeared to permit supersonic flight over land with acceptable sonic-boom characteristics.

In 2002 DARPA awarded Northrop Grumman a contract to design a long-range supersonic cruise aircraft that would operate with a less intense sonic boom, including variants for a long-range military strike aircraft and a civil business jet (designed in conjunction with Raytheon). DARPA then initiated a Shaped Sonic Boom Demonstration (SSBD) Program with Northrop Grumman, including major participation from NASA

Langley and Dryden, Naval Air Systems Command, Lockheed Martin, General Electric, Boeing, Gulfstream, Wyle Laboratories, and Eagle Aeronautics. The DARPA program's objective was to demonstrate that the far-field sonic-boom shock overpressure distribution for a configuration with a long forebody could be substantially altered in a favorable manner by geometric modifications. Northrop Grumman modified an F-5E fighter aircraft for the demonstration by installing a specially shaped "nose glove" substructure and a composite skin to the underside of the fuselage.

History was made on August 27, 2003, when the modified F-5E aircraft flew through a Dryden test range at supersonic speeds and demonstrated the first-ever in-flight exhibition of boom shaping. Ground-based and flight instrumentation measured the shape and magnitude of the sonic boom's atmospheric characteristics. A NASA F-15B also flew behind and below the F-5E to measure that aircraft's near-field shockwave patterns at a speed of about Mach 1.4. An unmodified F-5E was also flown to provide data for comparison to demonstrate the effect of aircraft shaping on sonic booms. An assessment of the sonic-boom overpressure measurements for each aircraft clearly shows the predicted reduced magnitude of the initial pressure rise for the modified F-5E.

This flight project made history as the first demonstration of the persistence of a shaped sonic-boom waveform, but more importantly, it proved numerous critics of the technology wrong, and has now given confidence to the potential for tailoring of sonic-boom characteristics.

Although the outlook for a large intercontinental commercial supersonic transport in the near future is clearly very pessimistic, an undertow of activities on supersonic technology and aircraft configurations has been forthcoming from the business-jet community. Supersonic business jets (SBJ) have frequently been discussed since the days of the NASA SCR Program, stimulated by the perspective that technical issues such as sonic-boom restrictions and airport noise requirements might be more easily met by the smaller SBJ. In addition, many believe that the SBJ passenger market will center on high-income individuals and companies for which saving time will be worth the operational costs.

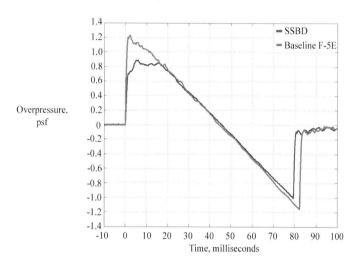

Data from a research flight of the modified F-5E compares the magnitude and character of the sonic boom overpressures for the demonstrator aircraft to those of an unmodified baseline F-5E aircraft. Note the significant reduction in the magnitude of the first overpressure wave. The demonstration finally gave the research community proof that the signature modification theory would work in the real world. (Graphic courtesy of NASA)

The highly modified Northrop-Grumman F-5E Shaped Sonic Boom Demonstration aircraft flies over Edwards Air Force Base in 2003. Note the reshaped nose and lower fuselage used to successfully demonstrate, for the first time, the ability to shape the far-field sonic-boom signature of an aircraft in supersonic flight. (Photo courtesy of NASA)

A modified NASA-Dryden F-15B flies with the extendable Gulfstream Quiet Spike concept in September 2006. The multi-segment device is designed to break the strong shock wave created at the nose of an aircraft flying at supersonic speeds into a series of weaker shocks, thereby decreasing the severity of the classic "N"-shaped sonic-boom signature. Extensive design efforts by Gulfstream included sonic-boom signature measurements of a model in the Langley Unitary Plan Wind Tunnel at Mach 1.8. (Photo courtesy of NASA)

Several collaborative SBJ ventures surfaced in the 1980s, including an effort involving Gulfstream and Sukhoi. At the National Business Aircraft Association (NBAA) convention in 2004, the Aerion Corp. and Supersonic Aerospace International (SAI) announced plans to launch full-scale development of supersonic business jets with cruise speeds of Mach 1.6 to 1.8.

Gulfstream has aggressively continued to pursue technologies relevant to supersonic business jets. The company supports most manufacturers' philosophy that an economically feasible SBJ must be capable of supersonic flight over land with acceptable sonic-boom signatures. In 2006 the company collaborated with NASA-Dryden to flight test a revolutionary Gulfstream concept for a sonic-boom mitigator. Designed and patented by Gulfstream, the "Quiet Spike" is an extendable, multi-segmented nose configuration designed to reduce the far-field effects of sonic booms. In principle, the concept breaks the strong bow shock of the aircraft into a series of weaker shocks in the far field. Wind-tunnel data of initial shock overpressures in the sonic-boom signature were successfully obtained during tests in the NASA Langley Unitary Plan Wind Tunnel at a Mach number of 1.8 to validate the concept at sub-scale and to correlate with theoretical predictions. The data indicated that the abrupt "N" shape of a typical aircraft signature can be altered to a much softer, more acceptable signature shape (more like a sine wave). In flight tests at Dryden, the nose of a NASA F-15B aircraft was modified with a Quiet Spike test article, which could be extended from 14 feet in subsonic flight to 24 feet in supersonic flight. To date the modified F-15B has flown the device to speeds of Mach 1.2, and the boom was extended and retracted successfully.

Based on the state of the art and the extensive history of aborted attempts to introduce supersonic transports, it has become clear that the most probable next-generation civil supersonic transport will be a business jet that can overfly land with acceptable sonic-boom and airport-noise characteristics. NASA Langley has historically been a major participant in the quest to achieve viable supersonic civil flight, and it is gratifying to observe that enterprising and innovative manufacturers of advanced business aircraft are leading the aeronautical community in this effort.

The aggressive efforts of the business-jet manufacturers and the recognition that numerous technical issues are more likely to be resolved for smaller supersonic aircraft almost assure the aeronautical world that the next generation of supersonic civil aircraft will be business jets. (Graphic courtesy of NASA)

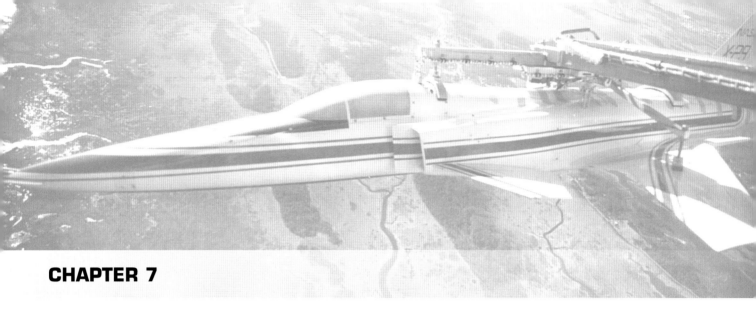

CHAPTER 7

MILITARY CONCEPTS

NASA inherited the proud tradition of the NACA in supporting the nation's military air forces in the conceptualization and development of new aircraft. The relationship of NASA and the military has been unique, and differs considerably from NASA's affiliation with the civil aeronautics industry. In serving its civil customers, NASA provides fundamental design information, consultative services, facilities, and cooperative studies on topics of mutual interest. However, NASA is usually not involved in industry assessments of potential aircraft designs or decisions to proceed into development of advanced aircraft concepts. Thus, industry selects specific data, expertise, facility usage, and other elements that are offered by NASA, but does not necessarily maintain communication or decision-making links with the agency during aircraft-related decisions.

In contrast to the civil industry situation, NASA assumes its role of a government agency in common activities with the military services. NASA basically contributes the same type of services for the military aeronautics community, with the important distinction that the agency frequently becomes a day-to-day partner with DoD on new aircraft projects. Historically, the NASA/DoD activities have been characterized by a strong interrelationship that features close and frequent planning, analysis, and mutual contributions that result in invaluable benefits to both parties. In recent times the cooperative ventures usually began with a recognized military mission or requirement that sparked interest in the availability and development of advanced technology within the nation's scientific community. On many occasions NASA has been requested to provide data on advanced aeronautical concepts for military decision-makers, and provide expertise for evaluations of proposed aircraft designs. In the development stages, NASA personnel conduct tests and studies within specific areas of expertise using unique facilities resulting in recommendations for modifications or endorsement of potential advantages for the design of interest. Finally, NASA researchers may be invited to participate in actual flight-test programs and the daily decisions that lead to the production and implementation within the military aircraft fleet. Once the aircraft is in operational status, changes in mission or unforeseen problems experienced by the services can result in request for support from NASA.

The NASA-military relationship for aeronautics has been an on-going characteristic of the programs conducted at NASA's research centers including Langley, Dryden, Ames, and Glenn. At Langley, notable projects have supported the Air Force, Navy, and Army for fixed-and rotary-wing vehicles that have contributed directly to the current military aircraft inventory. The following examples are indicative of typical projects of advanced military aircraft designs that did not proceed to production.

General Dynamics F-16XL

In the mid-1970s, Langley's aeronautical research staff were recognized leaders in the development of databases and design methods for efficient supersonic configurations. This expertise had developed as a result of participation in previous national and NASA civil supersonic transport programs as well as basic research on efficient supersonic designs with highly swept wings. Members of the Fort Worth Division of General Dynamics (now Lockheed Martin) approached Langley researchers and briefed them on company interests in developing a new wing configuration for the F-16 that would permit efficient supersonic performance with comparable transonic performance to the existing F-16. The new swept wing would be inserted on the existing fuselage and structure of the original configuration, and modified using fuselage inserts. The development of this advanced version of the F-16 offered an opportunity for NASA researchers to apply their skills and tools in a real-time technology-transfer arrangement.

In 1977, Langley and General Dynamics initiated a cooperative study to design a new "cranked" arrow wing for the F-16. Personnel from General Dynamics arrived at Langley for temporary duty and worked directly with the NASA staff. The study progressed into the fabrication and testing of several wind-tunnel models, and the results of supersonic- and transonic-tunnel model tests indicated that a viable wing could be designed. With these favorable results in hand, General Dynamics

In addition to supersonic and transonic activities in support of the F-16XL, Langley contributed wind-tunnel investigations in other specialized areas. This photograph shows a spin-tunnel model of the F-16XL configuration in early 1980. Note the deflected outer wing leading-edge flaps that were proposed as both high-lift devices and as roll-control surfaces. (Photo courtesy of NASA)

Langley aerodynamicists joined a team from General Dynamics to design an efficient supersonic wing for the F-16. Using design methods eveloped and validated by Langley in its supersonic transport studies, the participants defined several candidate wing planforms that would meet supersonic cruise requirements. One of the study requirements was to maintain a transonic maneuver capability equal to or better than the baseline F-16. Shown here in the Langley 7- by 10-Foot High Speed Tunnel in 1978 is a model used to provide validation data for a computational study of the impact of camber and twist on transonic performance. The study's results proved that a supersonic-wing design could have outstanding transonic maneuverability and influenced the final shaping of the wing of the F-16XL prototypes. (Photo courtesy of NASA)

One of Langley's areas of expertise is aeroelasticity, including the prediction and control of airframe flutter. As the F-16XL program progressed, the configuration became a candidate for an advanced strike fighter for the Air Force and was redesignated the F-16E. This photograph shows a large model of the F-16E mounted in the Langley 16-Foot Transonic Dynamics Tunnel in late 1981 during flutter-clearance tests. Note the instrumentation on the vertical tail used to measure dynamic tail loads. (Photo courtesy of NASA)

initiated the company-funded development of an F-16 derivative with supersonic cruise capability that retained the conventional F-16's transonic maneuverability. A cooperative program was initiated with Langley to participate in the development of what was initially called the Supersonic Cruise and Maneuver Prototype (SCAMP).

At this point, the scope of Langley's involvement broadened considerably. Testing of the SCAMP configuration took place in several facilities, including the Langley Unitary Plan Wind Tunnel to study supersonic performance, the Langley 7- by 10-Foot High-Speed Tunnel to study transonic performance, the Langley 16-Foot Transonic Dynamics Tunnel to study inlet integration, the Langley Full Scale Tunnel to study dynamic stability and control, the Langley Differential Maneuvering Simulator to assess handling qualities, the Langley 20-Foot Spin Tunnel to evaluate spin and recovery characteristics, and a helicopter drop model to study spin-entry behavior.

Although the cranked arrow wing planform derived in the cooperative study was highly effective in increasing the supersonic performance of the basic F-16, the projected low-speed characteristics and weight-and-balance considerations caused modifications to the configuration during the development process. The wingspan and sweep back were varied in attempts to resolve low-speed flight issues, and the NASA/General Dynamics team arrived at fixes that included modifying the wing apex with a rounded planform.

Research on the SCAMP configuration covered other areas of interest to Langley researchers, including the application of vortex flaps on the highly swept leading edge for improved low-speed and transonic performance, automatic spin-prevention concepts, and optimized concepts for carriage of external stores at supersonic speeds. The final configuration became known as the F-16XL (later designated the F-16E), which displayed an excellent combination of reduced supersonic wave drag, vortex lift for transonic and low-speed maneuvers, low structural weight, and good transonic performance.

As the F-16XL configuration matured, flutter testing was also conducted in the Langley 16-Foot Transonic Dynamics Tunnel to clear the airplane design for flight. Subsequently, two F-16XL demonstrator aircraft (a one-seat and a two-seat) were built from modified F-16 aircraft and entered flight tests in mid-1982. Highly

zThe two-place F-16XL prototype displays several features derived from inputs by Langley researchers, including the advanced supersonic-wing configuration, the reflexed wing apex modification, and outer-wing upper-surface fences used to enhance stability. During the F-16XL's development, Air Force interest shifted from an efficient supersonic-cruise version of the F-16 to a multi-role ground-attack fighter. After flight evaluations of prototype competitors, the Air Force selected the F-15E for development, and the two F-16XL prototypes were subsequently used in NASA flight-research programs. (Photo courtesy of NASA)

successful flight tests of the aircraft by General Dynamics and the Air Force at Edwards Air Force Base confirmed the accuracy of the wing-design procedures, wind-tunnel predictions, and control-system designs that had resulted from the cooperative program. Unfortunately, initial Air Force interest in a supersonic-cruise version of the F-16 diminished when the development of a dual-role fighter with ground-strike capability became a priority. Although the F-16XL could efficiently carry a significant number of weapons for ground-strike missions, the Air Force ultimately selected the F-15E in 1983 for developmental funding and terminated its interest in the F-16XL. After termination of the Air Force program, NASA obtained the two F-16XL aircraft for studies of supersonic laminar flow control stimulated by the potential application of the technology to future supersonic transports. Langley and Dryden partnered with industry teams in the effort, which included flight tests at Dryden in the

NASA used the F-16XL prototypes for flight-research efforts in supersonic laminar flow, sonic-boom characterization, and detailed experimental and computational studies of the vortex flow over the wing. Shown here is the single-seat prototype during the Cranked-Arrow Wing Aerodynamic Project at NASA-Dryden in 1996. The aircraft's upper surfaces were painted black as an aid in flow-visualization studies; tufts and pressure-measuring instrumentation were used to characterize the flow phenomena on the wing. The photograph was taken for an angle of attack of 21 degrees at subsonic flight conditions. Note the angularities of the tufts near the leading edge of the wing, indicative of vortical flow. Results of the flight tests were correlated with computational predictions and used to advance the state of the art in design methods for highly swept wings. (Photo courtesy of NASA)

Engineer Sue Grafton inspects a large 0.15-scale free-flight model of the F-16XL tested in the NASA Langley Full Scale Tunnel. The original "SCAMP" configuration was modified with the reflexed wing-apex configuration shown here to eliminate a pitch-up tendency at moderate angles of attack. This modification resulted in outstanding flight characteristics during the Langley free-flight evaluation and was incorporated in the final design of the full-scale aircraft. (Photo courtesy of NASA)

1990s. The aircraft were also used in sonic-boom studies and detailed analyses of the vortical flow over the cranked arrow wing.

Grumman X-29

In the early 1980s, progress in advanced composite construction methodology sparked a new interest in an old aeronautical concept: forward-swept wings. During World War II, the Germans had initiated the development of a forward-swept wing (FSW) bomber, built by Junkers and designated the Ju 287. The FSW concept was based on the theory that the onset of adverse compressibility effects could be delayed at high subsonic or transonic speeds by sweeping the wing either backward or forward. Early research had shown that sweeping the wing rearward resulted in several undesirable characteristics caused by airflow separation at the wingtips at moderate angles of attack. The problems included uncontrollable longitudinal instability (pitch up) and loss of lateral control provided by conventional ailerons at high angles of attack. Unfortunately, flight tests of the Ju 287 revealed unacceptable increases in g-loading caused by structural deformation (structural divergence) of the aluminum forward-swept wing structure. In view of this problem, interest in FSW configurations evaporated following World War II.

In the 1970s, however, DARPA's Dr. Norris J. Krone Jr. conducted research on the possibility of eliminating structural divergence for FSW airplanes by using advanced tailored composite materials. Such materials promised light weight and significantly increased structural rigidity compared to aluminum. With the advocacy of Krone, DARPA sought industry proposals for FSW research aircraft in 1977 with responses forthcoming from Grumman, Rockwell, and General Dynamics. Following assessments of the candidate configurations, Grumman was awarded a contract to develop an FSW demonstrator aircraft designated X-29 in 1981.

To support the development of the X-29, DARPA requested the assistance of Langley researchers in the areas of aeroelasticity, spin recovery, and flight dynamics at high angles of attack. Langley personnel and facilities had participated in obtaining data during the DARPA competitive selection process, obtaining information on configurations submitted by the three industry competitors. Activities at Langley in support of the Grumman X-29 included aeroelastic and flutter tests in the 16-foot transonic dynamics tunnel, free-flight model tests in the full-scale tunnel, spin and spin-recovery tests in the spin tunnel, pilot evaluations of the handling qualities of the X-29 demonstrator in the Langley Differential Maneuvering Simulator (DMS), and evaluation of spin-entry characteristics using a helicopter drop model.

During the free-flight model tests in the full-scale tunnel, a major surprise occurred when the model exhibited large uncommanded rolling motions at high angles of attack, suggestive of loss of lift on the outboard wing panels. Such a result was totally unexpected based on the ability of a forward swept wing to maintain airflow over its wingtips. Additional wind-tunnel tests were immediately undertaken to determine the cause of the discrepancy. The test team found that the undesirable rolling motions were caused by the interaction of vortical flows shed by the long pointed nose of the X-29 with air flows over the aircraft's fuselage and wing. The flow interactions were so strong that they resulted in undesirable rolling motions, despite the fact that the wingtip flows were attached as expected. These results proved to be especially valuable in the flight-control system's design, which eliminated the problem for the X-29 airplane.

Langley researchers also conducted pioneering work with the X-29 free-flight model control system. The X-29 airplane was designed with a very large aerodynamic longitudinal instability (relaxed static stability) that was stabilized by its redundant digital flight-control system. The level of relaxed static stability for the X-29 airplane was an order of magnitude more unstable than that for the F-16 fighter. During the wind-tunnel free-flight tests, Langley staff devised a similar feedback system for the model that permitted satisfactory flight.

Results from the DMS pilot simulator studies aided in the design of the flight-control system as well as provided pilots with a preview of the anticipated flying qualities of the aircraft before flight tests of the full-scale airplane began at NASA Dryden. Data from the simulator activities provided control law concepts for coordinated roll maneuvers at high angles of attack, suppression of the rolling motions discussed earlier, and options for automatic spin prevention.

Meanwhile, Langley researchers in the spin tunnel evaluated the spin and spin-recovery characteristics of an X-29 model and determined the size of emergency

Langley researchers in the full-scale tunnel joined in a cooperative NASA/DARPA project to support development of a forward-swept wing flight demonstrator sponsored by DARPA. In 1978, competing configurations from Rockwell, General Dynamics, and Grumman were tested in the Langley 12-Foot Low-Speed Tunnel and the Langley Full Scale Tunnel. Project Engineer Sue Grafton is shown with the General Dynamics forward-swept-wing version of the F-16. The model was used for conventional tunnel tests to measure aerodynamic performance, stability, and control for high-angle-of-attack conditions. Results of the brief test program were very positive. (Photo courtesy of NASA)

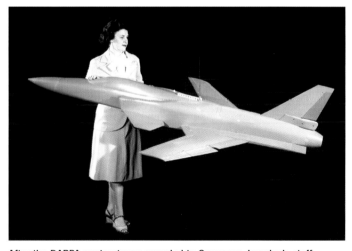

After the DARPA contract was awarded to Grumman, Langley's staff fabricated models of the X-29 for free-flight tests, spin-tunnel tests, and outdoor drop-model tests. This 0.16-scale free-flight model was studied in the full-scale tunnel in 1980. Despite having the highest level of aerodynamic longitudinal instability ever flown for a free-flight model, the augmented configuration had superb longitudinal characteristics. With the roll-stabilization system off, large roll oscillations were encountered; however, the augmented (basic) configuration had outstanding behavior. (Photo courtesy of NASA)

Researcher Rodney H. Ricketts poses with a large half-model of a Grumman configuration used in the 16-Foot Transonic Dynamics Tunnel for aeroelastic studies in 1979. Specific objectives were to characterize the structural divergence tendencies of the forward-swept wing and correlate the results with design methods. (Photo courtesy of NASA)

Free-flight model of the Grumman X-29 flying at high angle of attack in the full-scale tunnel. Note the nose boom vanes used to measure angle of attack and angle of sideslip for feedback to stabilization systems, the deflected surfaces on the aft fuselage used for trim, and the nose-down deflection of the canard. For this aerodynamically unstable design, nose-down trim was required at high angles of attack, rather than the nose-up setting required for stable configurations. (Photo courtesy of NASA)

spin-recovery parachute required for the flight-test program. Results of the drop model program correlated well with the free-flight model and spin-tunnel model results, and extended the knowledge base into the post-stall region. As an integrated source of information, these model and simulator results provided the X-29 flight-test team with an invaluable database to begin flight tests of the full-scale aircraft.

One of the X-29 program's major objectives was to demonstrate composites' ability to mitigate the wing-divergence tendencies of FSW configurations. Langley experts in aeroelasticity were major participants in pre-flight analyses of the aircraft. In 1983, the staff of the Langley 16-Foot Transonic Dynamics Tunnel conducted tests to develop techniques for analyzing various structural modes and flutter tendencies for the X-29.

Two X-29 demonstrator aircraft flew at Dryden from 1984 to 1992 in a series of phased investigations with different objectives. In the first phase, the X-29 demonstrated the ability of aeroelastic tailoring to prevent wing divergence, and the ability of its digital flight-control system to provide good handling qualities despite a high level of aerodynamic longitudinal instability. In the second flight program, the tests focused on high angle-of-attack maneuverability and control. The airplane, which depended solely on aerodynamic controls (no thrust vectoring) performed admirably, executing roll maneuvers at angles of attack as high as 45 degrees without tendency to depart from controlled flight. The very successful demonstration of advanced technologies with the X-29 was in no small part due to the hard work and background data obtained by the NASA/Grumman/Air Force team prior to flight test.

Grumman X-29 in flight in 1985 above mountain ranges near NASA Dryden Flight Research Center. The flight-demonstration program's results provided a wealth of design information on tailored composite structures, variable-camber wings, forward-swept wings, and advanced digital flight-control systems. Two demonstrator aircraft conducted 374 flights in the very productive program. This view shows the radical forward-swept wing to advantage as well as the close-coupled canard. Nose strakes used to enhance directional stability are mounted at the base of the nose boom. (Photo courtesy of NASA)

Although the spin-tunnel facility provides information on the developed spin and spin recovery, and the free-flight technique used in the full-scale tunnel indicates the stability of a configuration at stall, neither can be used to evaluate the tendency of an aircraft to enter a spin (spin entry). Langley developed an outdoor drop-model technique for this purpose. Shown here is a large-scale model of the X-29 mounted on a model-drop apparatus attached to the side of a helicopter during a typical spin-entry evaluation. The model was released at an altitude of about 3,000 feet over Plum Tree Island, an abandoned bombing range near Langley. Ground-based operators controlled the model, inputting various control manipulations in a deliberate attempt to enter a spin. At the termination of the test, the model was recovered with an onboard parachute. (Photo courtesy of NASA)

Following the primary X-29 research flight program, the Air Force, Grumman, and Dryden conducted a series of flight tests to evaluate the use of vortex control on the X-29 forebody for control augmentation at high angles of attack. Today one of the aircraft is on display at the National Museum of the United States Air Force at Wright-Patterson Air Force Base, and the other is on display at the Dryden Flight Research Center.

NASA F/A-18 High Alpha Research Vehicle (HARV)

During the Vietnam War, U.S. military forces were equipped with fighter aircraft that had been designed without great concern on maneuverability at high angles-of-attack. Aerial engagements with the enemy, however, showed that the enemy forces employed highly maneuverable aircraft that forced U.S. pilots to push their aircraft into using high angles-of-attack to generate the lift required for maneuvers. Unfortunately, aircraft such as the F-4 Phantom and the A-7 Corsair II became directionally unstable at high angles-of-attack, resulting in an abrupt loss of control and potential spin entry. Many of these aircraft were lost in accidents involving poor behavior in this flight regime. In the mid-1970s, a national emphasis on providing a new generation of fighter aircraft with excellent high angle-of-attack char-

acteristics emerged, and NASA was an active participant in providing the research and design data required for the task. With several cooperative industry partners, the agency contributed support to the development of today's F-16, F/A-18, and F-22, all of which have outstanding maneuverability and spin resistance.

During the early 1980s, NASA researchers at Langley, Dryden, and Ames conceived and developed a research project known as the NASA High Angle of Attack Program (HATP). The program included an assessment and development of computational and experimental aerodynamic methods, development of concepts for control augmentation at high angles-of-attack, and evaluations of handling qualities. The HATP planners specifically designed a series of coordinated efforts involving wind-tunnel tests, computer codes, piloted simulator studies, and aircraft flight tests. After evaluating the known advantages and disadvantages of available aircraft to serve as the configuration of interest for correlation of the research efforts, the F/A-18 was chosen as the baseline for the program. The aircraft would be equipped with special instrumentation to measure and analyze airflow properties, a research flight-control system, and concepts to improve controllability at high angles-of-attack, including thrust vectoring and forebody flow-control surfaces. One of the prototype F/A-18A aircraft that had been used for spin testing by McDonnell Douglas and the Navy in the F/A-18 development program was acquired on loan, refurbished, and modified for flight testing at Dryden. The F/A-18 research aircraft was known as the High Alpha Research Vehicle (HARV).

At Langley, selection of the F/A-18 as the baseline configuration resulted in extensive wind-tunnel and computational studies of aerodynamics at extreme angles-of-attack in an effort to provide design information and tools for future aircraft. Pressure distributions over the configuration were measured in several wind tunnels using highly instrumented models, and data were analyzed and prepared for correlation with results obtained from subsequent flight tests with the full-scale aircraft at Dryden. Computational fluid dynamics studies for similar flight conditions were used to correlate the wind-tunnel and flight-test results. The emergence of powerful new computational methods and supercomputers provided for an assessment and validation of these new tools. One of the HATP activities' most unique contributions was aerodynamic data measured for similar conditions and instrumentation between computational, experimental, and flight sources.

In the HATP's first phase, flight-testing of the HARV for aerodynamic studies began in April 1987. First-ever flight data for visualization of flow over the aircraft's fuselage forebody surfaces using colored dyes was accomplished, visualization of the strong vortex flow from the wing-fuselage leading-edge extension (LEX) was conducted using smoke, and wool tufts were used to visualize flow over the wing surfaces at high angles of attack. The data gathered from these unique flight tests contributed directly to the validation of computational methodology, wind-tunnel/flight correlation procedures, and understanding of fundamental fluid physics ranging from vortex breakdown to vertical tail buffet prediction.

Researchers at Langley had, for several years, led the American effort in thrust-vectoring concepts, including demonstrations of the effectiveness of control augmentation provided by thrust vectoring for high angle-of-attack flight conditions. In addition to directional instability produced by the effects of airflow separation, a major problem is the loss of lateral-directional control power by conventional aerodynamic control surfaces at high angles-of-attack. This phenomenon occurs because of the impingement of low-energy separated airflow wakes on the control surfaces during stalled flight. Langley proposed that two different concepts be explored to provide control augmentation. These concepts were engine thrust vectoring and forebody vortex flow control.

Initially, a series of powered free-flight model tests in the full-scale tunnel were conducted to demonstrate the dramatic increase in controllability provided by thrust vectoring. Using representative models of the F/A-18 with thrust vectoring provided by compressed air simulating engine thrust, researchers demonstrated the ability to fly an F/A-18 model to extreme angles-of-attack (beyond 75 degrees) in level flight without loss of control. Results of the tests indicated that relatively

The NASA F/A-18 High Alpha Research Vehicle (HARV) in flight at the NASA Dryden Flight Research Center. As the centerpiece of a multi-discipline research program, the HARV provided extensive experience with thrust-vectoring technologies, aerodynamic data for correlation with wind-tunnel tests, and validation of computational aerodynamic-prediction methods. In a three-phase program, the aircraft was first used in studies of aerodynamics at high angles of attack. In the second phase, the vehicle was modified with a research thrust-vectoring concept and associated control laws. Used in handling-quality studies and flight at extreme angles of attack, the HARV provided data of great use to the design community. In the third series of studies, the HARV was modified with nose-mounted strakes for directional control augmentation at high angles of attack. (Photo courtesy of NASA)

Rear view of the F/A-18 HARV during ground tests of its thrust-vectoring system dramatically illustrates the deflection of the engine exhaust by the deflected vanes. This unique capability provided NASA researchers with an unprecedented opportunity to analyze the structural loads and activation of vectoring concepts, develop control-system design methods for vectoring, and demonstrate the tactical advantages provided by thrust vectoring. The powerful control provided by vectoring also was used to stabilize the HARV at extreme angles of attack for aerodynamic studies. (Photo courtesy of NASA)

The philosophy for modifying the HARV with a thrust-vectoring system focused on a relatively low-cost approach, using the existing engines of the F/A-18. Langley researchers had conducted wind-tunnel tests of simple paddle-vanes in the engine exhaust as a candidate approach for the full-scale HARV airplane. Shown here is the vectoring concept implemented on the free-flight model of the HARV. Note the three-vane arrangement of vectoring surfaces used to provide both pitch and yaw control for each engine nozzle. The afterburner of the full-scale F/A-18 aircraft was removed for the vectoring system to minimize loads on the vectoring structure, and the actuators and vectoring structure were mounted externally to the existing airframe. The box above the vectoring vanes is a model of the emergency spin-recovery parachute container carried on the full-scale aircraft, and the parachute-attachment structure is immediately below the container. (Photo courtesy of NASA)

simple thrust-turning concepts such as exhaust paddles could be used very effectively for this purpose. With these results in hand, researchers decided to implement a rudimentary three-paddle thrust-vectoring system on the full-scale F/A-18 at Dryden for flight evaluations and demonstrations. McDonnell Douglas was awarded a contract to modify the HARV aircraft with a paddle-based mechanical thrust-vectoring system along with a special research flight-control system to accommodate the vectoring capability. The staffs of Dryden and Langley also made major contributions to this effort.

In 1991, the HARV thrust vectoring had become operational, introducing the second major phase of the HATP in which thrust-vectoring capability was demonstrated and evaluated. The direct application of thrust vectoring was in two areas. In the first application, the powerful controls provided by vectoring were used to stabilize the aircraft at extreme angles of attack for aerodynamic studies. The second application of vectoring was to demonstrate the improved low-speed, high-angle-of-attack maneuverability predicted by the Langley model testing and simulator studies.

Researchers at Langley had pursued the use of actuated nose strakes to control powerful vortices shed off the long pointed noses of current military fighters. The engineering community had long known that strong vortex flow fields emanate from long pointed bodies at high angles of attack. In fact, several operational aircraft had exhibited asymmetric shedding of such vortices, resulting in an uncontrollable yaw motion. Researchers theorized that forcing the separation of vortices from one side of the nose in a controlled manner might create very large yaw control for conditions in which the level of yaw control from conventional rudders typically diminishes. Extensive conventional wind-tunnel testing at subsonic and transonic speeds in several wind tunnels was accomplished with F/A-18 models equipped with deflectable nose strakes to augment yaw control at high angles-of-attack. A Langley-Dryden team was subsequently formed to design and implement this concept on the HARV aircraft.

In 1995, the HARV nose-strake flight investigations at Dryden began under a project known as Actuated Nose Strakes for Enhanced Rolling (ANSER). In flight tests that continued into 1996, the operational utility on the powerful yaw control provided by the strakes were evaluated and demonstrated both individually and in combination with the aircraft's thrust-vectoring system capability.

As part of a large NASA effort and accelerating the state of the art in high angle-of-attack technology, the HARV research aircraft was a centerpiece for the program. The vehicle completed 385 research flights with angles-of-attack up to 70 degrees, vividly demonstrating the potential benefits of thrust vectoring and

The aeronautics community has long been aware of the fact that long, pointed noses shed powerful vortex flows at high angles of attack. With advanced flight-control systems routinely used today, researchers pursued the concept of deflectable nose strakes to generate vortex control to augment yaw control at high angles of attack. Wind-tunnel tests of generic models and computational fluid dynamics results such as the one shown here were used to configure a system for evaluation during flight tests of the HARV. (Graphic courtesy of NASA)

Shown on the flight ramp at Dryden in 1995, the HARV displays the hinged, fold-out strakes used for yaw control at high angles of attack. In the Actuated Nose Strakes for Enhanced Rolling (ANSER) project, a HARV radome was modified with the strakes in Langley fabrication shops. The modified radome was then transferred to Dryden, where it was outfitted with instrumentation, flight hardware, and software by Dryden's staff and prepared for flight. (Photo courtesy of NASA)

In 1988, Langley researchers evaluated the effectiveness of vane-type thrust-vectoring concepts and forebody control strakes with a series of conventional and free-flight model tests in the NASA Langley Full Scale Tunnel. In this photograph, the free-flight model has been modified with a rudimentary thrust-vector concept and deflectable forebody strakes. In addition to wind-tunnel and computational aerodynamic studies, piloted simulator evaluations were conducted to develop control laws required for the new control concept. (Photo courtesy of NASA)

During flight evaluations, Dryden and Langley test pilots found the forebody strakes to be powerful yaw-control devices at high angles-of-attack. The strong vortical flow acting on the long forebody of the HARV resulted in very high levels of yaw control in conditions in which conventional rudders typically lose effectiveness because of immersion of the vertical tails in low-energy flow from stalled wings. (Photo courtesy of NASA)

forebody flow control. In addition, the maturation and validation of experimental and computational aerodynamic-prediction capabilities was significantly enhanced and disseminated to industry for future high-performance aircraft program projects.

Rockwell X-31

In the 1980s, international interest increased in the design of highly maneuverable fighter aircraft capable of unprecedented flight at extreme angles-of-attack and low airspeed. In addition to advances in digital flight-control system technologies, this interest was spurred by the emergence of thrust vectoring for additional control power at high angles of attack.

At high angles of attack and low-speed conditions, conventional fighter aircraft typically exhibit a loss of control authority from aerodynamic surfaces such as ailerons, elevators, and rudders. In addition, conventional configurations may exhibit abrupt, severe aerodynamic instability in yaw or roll, resulting in inadvertent loss of control and to spin entry. NASA Langley had been an active player in the research and development activities for solutions to these deficiencies, and was poised to participate in the development of radical new fighter configurations for this mission.

In 1984, Langley researchers joined with Rockwell (now Boeing) engineers to study thrust-vectoring fighter configurations capable of agile maneuvers at high angles of attack. In a series of investigations in the full scale tunnel at Langley, the research team evolved candidate configurations and provided considerable data useful in the design of full-scale aircraft. These exploratory studies vividly demonstrated the ability of vectoring engine thrust to provide large control moments for

"care-free" maneuvers at high angles of attack. A critical factor in this technology was Langley's pioneering work in developing thrust-vectoring configurations adaptable to fighter configurations.

In 1986, DARPA initiated the first International (U.S. and Federal Republic of West Germany) X-Plane Program to demonstrate the value of using thrust vectoring with advanced digital control systems to provide unprecedented maneuverability at very high angles of attack. The demonstrator aircraft was designated X-31, and program participants included Rockwell, Messerschmitt-Bolkow-Blohm (MBB), and NASA. The program became known as the X-31 Enhanced Fighter Maneuverability (EFM) Program. The aircraft used a relatively low-cost thrust-vectoring system based on the use of three movable graphite-epoxy paddles located around

Free-flight model of the Rockwell X-31 flying in the NASA Langley Full Scale Tunnel in 1990. Langley facilities contributed extensive developmental data and analysis to the X-31 project team in the areas of thrust-vectoring vane-configuration effects, spin and recovery characteristics, post-stall gyrations and recovery techniques, and control-system design options. (Photo courtesy of NASA)

the engine exhaust, with various combinations of paddle deflections to provide pitch- and yaw-control augmentation.

Beginning in 1987, Langley researchers conducted extensive studies of the stability, control, and thrust-vectoring system of the X-31 configuration. Activities included tests using several X-31 models in the full-scale tunnel, the spin tunnel, the 12-foot low-speed tunnel, the 14- by 22-foot tunnel, the 16-foot transonic tunnel, a jet exit test facility, a radio-controlled drop model, and in piloted simulators. Especially valuable results of free-flight tests of an X-31 model during the tests in the full-scale tunnel provided information on lateral stability that contributed directly to the design of the flight-control system. Likewise, critical data were provided on the post-stall gyrations and controllability of the configuration during the drop model tests. In addition, final design information on the effectiveness of various configurations of the thrust-vectoring paddles used by the X-31 was provided by data from over 500 paddle and nozzle configurations in the jet exit test facility.

A Langley technician in the launch helicopter looks over the status of a radio-controlled drop model of the X-31. The post-stall motions of the X-31 during controlled flight to extreme angles of attack were rapid and precise, providing the pilot of the full-scale aircraft with new maneuvering options during air combat. However, if the X-31 was intentionally forced into a spin entry, it exhibited severe, disorienting post-stall gyrations that required specific recovery-control inputs. This information, obtained from Langley's drop-model tests, was provided to the X-31 flight test team before the full-scale vehicle flew such maneuvers and was invaluable guidance for the flight program. Langley also maintained the drop-model capability for on-call requests for additional tests during the flight program at Dryden. (Photo courtesy of NASA)

Langley personnel also continued their participation in the X-31 program after NASA-Dryden became the responsible test organization during the flight-test phase. Testing of the X-31s began at Dryden in February 1992 and concluded in 1995. One of the X-31 aircraft was destroyed in a flight accident caused by air-data failures in January 1995.

During subsequent flight-evaluation tests at Dryden, Langley provided technical support and on two occasions provided rapid solutions to critical stability and control problems. The first problem encountered in flight tests was insufficient nose-down pitch control at extreme angles of attack, and the second problem was the existence of large asymmetric yawing moments that exceeded the available level of yaw control. In both cases, a request to Langley from the flight-test organization resulted in immediate wind-tunnel tests to provide potential solutions. Langley's researchers quickly identified airframe modifications consisting of strakes along the fuselage afterbody and nose tip. The Langley recommendations, which were given within a week of the test request, were adopted and evaluated in flight, providing a timely solution to the problem. During the Dryden-based program, the X-31 flew an X-plane record of more than 500 research flights in 52 months with 14 pilots from NASA, the U.S. Navy and Marine Corps, the U.S. Air Force, the German Air Force, Rockwell International, and Deutsche Aerospace.

In the 1990s, the U.S. Navy sponsored the Vectoring Extremely Short Takeoff and Landing Control and Tailless Operation Research (VECTOR) program using

The X-31 flies at high angle-of-attack at Dryden in 1994. The flight-test program vividly demonstrated the advantages of thrust vectoring for close-in air-combat maneuvers. Designed as an operational concept demonstrator, the X-31 could execute

radical minimum-radius turns by flying well beyond the stall limits of conventional aircraft. (Photo courtesy of NASA)

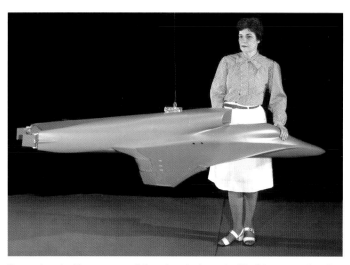

Spurred on by the success of thrust vectoring retrofitted to conventional aircraft, the Langley staff experimented with more-radical configurations that were designed from the onset for thrust vectoring. Here a radical tailless free-flight model of an in-house advanced fighter design is inspected prior to full-scale tunnel tests in 1985. Langley free-flight tests of models of the F/A-18, X-29, F-16 and others equipped with thrust vectoring had all shown no angle-of-attack limits, and researchers theorized that the tails could be removed for flight at high angles of attack if thrust vectoring was provided. Auxiliary tails or other concepts would be required for flight at low angles of attack and high speeds, or if engine thrust was loss. (Photo courtesy of NASA)

The tailless "thrust-vector-control" model flies at extreme angle of attack in the Langley Full Scale Tunnel in 1985. Researchers used a three-vane control system to provide pitch and yaw control, together with feedback signals from a nose boom for stability augmentation. In some tests, the tunnel airspeed was reduced and stopped as the model completed a completely controllable "tailsitter" transition to hovering flight. (Photo courtesy of NASA)

the remaining X-31 aircraft. In a joint venture between the Navy, Germany's federal defense procurement agency, European Aeronautic Defense and Space (EADS) Corp., and Boeing Aerospace, the flight-project objectives were initially demonstrations of extremely short takeoff and landing (ESTOL) capabilities using thrust vectoring, a flush air-data system, and reduced vertical tail configurations. A successful demonstration of ESTOL capabilities was of special interest to the Navy for carrier operations. Flight testing in the VECTOR program was led by the Naval Air Systems Command at Patuxent River, Maryland, from 2001 to 2003.

The X-31 VECTOR program concluded with the demonstration of a fully automated, high angle-of-attack landing. The use of thrust vectoring and advanced flight controls led to a 30-percent reduction in the landing speed of the X-31, using an approach angle of attack of 24 degrees compared to a normal angle of attack of 12 degrees. An advanced air-data system (no booms or vanes) provided a flight envelope up to a Mach number of 1.2 and 70 degrees angle-of-attack.

NASA F-106B Vortex Flap

In the 1970s, the emergence of high-performance fighter aircraft that used vortex-induced lift enhancement for increased maneuverability led to several cooperative research programs between NASA, DoD, and industry. At Langley, a cooperative study with Northrop on wing-fuselage strake configurations (leading-edge extensions) resulted in a large database on the effects of various strake configurations on aerodynamic performance and stability. Physical understanding of vortex flow formation, growth, and breakdown was significantly improved and available for inclusion in studies of advanced civil and military aircraft. This fundamental research data and the tools that were produced have been used in a myriad of follow-on programs including the development of today's advanced F/A-18 Super Hornet.

As previously discussed, when General Dynamics approached Langley researchers for cooperative studies of a re-winged F-16 variant known as SCAMP, a program goal was to develop a supersonic-efficient wing while retaining transonic maneuver capability comparable to the basic F-16. Providing the supersonic-cruise capability quickly focused the project participants to propose a relatively thin, cranked-arrow wing similar to those studied by NASA and industry for civil supersonic transport configurations. Unfortunately, highly swept arrow wings exhibit flow separation around the swept leading edge of the wing at moderate angles-of-attack, resulting in the formation of vortices that cause excessive drag and greatly diminished aerodynamic efficiency for maneuvering flight. Langley's experts in vortical-flow technologies accepted the challenge of identifying concepts that could be used to improve the transonic-maneuver capability of the new wing.

Armed with experimental experience on vortex flow control and emerging computational fluid dynamics capabilities, researchers began to examine the benefits of variations in wing camber (airfoil shape) for the new cranked-arrow wing design. After extensive computational analyses and coordinated wind-tunnel tests of modified SCAMP configurations, they began to visualize the optimal wing design for transonic maneuvering. This design effectively suppressed the generation of leading-edge vortices and dictated their paths along the wing surface when separation could not be avoided at high angles-of-attack. The resulting wing design did provide the desired level of transonic maneuverability. The more-efficient wing design, however, involved an extraordinary distribution of wing camber and shaping that produced serious concerns regarding the manufacturing and structural design of such wing shapes.

Driven by the desire to explore the relative effectiveness of less-optimum wing designs that might still provide vortex control, the Langley staff focused on a concept known as the "vortex flap" which uses a potentially simpler, specially designed, deflectable leading-edge flap to capture and control the wing leading-edge vortices shed by highly swept-wings. The vortex flap concept had been explored independently and cooperatively by NASA and industry in the late 1970s. In exploratory testing, researchers found that certain combinations of deflected full-span leading- and trailing-edge flaps on a planar (no camber) wing produced almost the same drag improvements at transonic speeds as a specially designed and transonically cambered wing. Initial assessments were conducted for applications ranging from civil supersonic transport configurations to the F-16XL.

In 1983, NASA undertook an aircraft flight project to assess the validity of the model predictions at full-scale conditions and other factors such as the potential impact of vortex-flap effects on aircraft-handling qual-

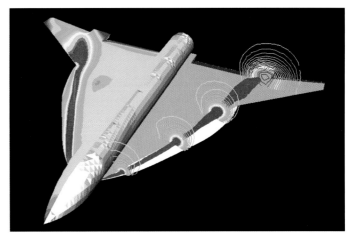

Computational analysis and wind-tunnel experiments were used in the 1970s to develop concepts to control the vortical flows generated at the leading edges of highly swept wings. The objective of the study was to develop more-efficient configurations for transonic and subsonic flight conditions. This computer-generated graphic shows flow conditions on the left and right wing panels of a cranked-arrow wing at moderate angles of attack. The aircraft's right wing panel displays the surface pressure, with red signifying a low-pressure region. The interaction of a strong leading-edge vortex has created a strong negative pressure along the leading edge. The aircraft's left wing panel shows the clearly defined vortex flow as it flows rearward, tornado-style, over the wing. (Graphic courtesy of NASA)

ities. For example, one concern expressed by the engineering community was whether dynamic aircraft rolling motions would cause disruption of the vortex-control capability of the vortex flap. With an F-106B aircraft in its inventory, Langley was poised to begin a flight project to provide answers to these and other questions. A series of sub-scale wind-tunnel tests from subsonic through transonic speeds was undertaken at Langley and Ames to provide aerodynamic data on a specific vortex-flap configuration that had been designed at Langley using Langley-developed computational codes. Free-flight model tests were conducted in the full-scale tunnel to evaluate effects on dynamic stability and control, and piloted simulator evaluations using the Langley Differential Maneuvering Simulator occurred. An in-house Langley team led the design and fabrication of the vortex flap. The flap was designed to be a "bolt on" configuration, fixed at an angle prior to flight and evaluated for several flap settings.

The F-106B was equipped with extensive instrumentation to measure pressures and visualize airflow over the wing in flight. A unique vapor-screen flow-visualization system was designed and implemented by the

Langley staff to permit analysis of the vortex flow fields over the wing in flight. This system "seeded" the airflow with a heated propylene glycol vapor pumped from the aircraft's missile bay and expelled through a probe under the left wing panel's leading edge. After a series of visualization test flights for the basic airplane in 1985, researchers conducted extensive analyses of the results in preparation for the vortex-flap investigation.

In 1988, the first flights of the F-106B vortex-flap experiments began a series of investigations that would last for about two years. Flights were conducted for vortex-flap angles of 30- and 40-degrees for Mach numbers from 0.3 to 0.9 and for altitudes up to 40,000 feet. The results of the flight study were very impressive. For example, the aircraft's sustained g capability was increased by about 28 percent at a Mach number of 0.7. In addition to the dramatic demonstration of improved performance, the flight project provided a valuable validation of the vortex-flap design process and experimental prediction capabilities.

Engineer Hal Baber inspects an F-106B vortex-flap model in 1984 prior to tests in the full-scale tunnel. The details of the geometry of the leading-edge flap were derived from extensive computational and tunnel studies over the speed range from takeoff and landing to transonic maneuvers. (Graphic courtesy of NASA)

Langley researchers developed an innovative flow-visualization technique for qualitative evaluations of the vortex flow structure over the wing in flight. Using a movable light sheet generated by an on-board system, the evaluators conducted night flights to observe the flow phenomena. For many conditions, the leading-edge vortex was seen to decompose into several smaller vortices. (Graphic courtesy of NASA)

Langley's F-106B vortex-flap research aircraft in flight in 1987. Note the flow-visualization equipment housing on the turtleback of the aircraft. Flow over the left wing panel was illuminated during flights at night. The right wing panel carried pressure instrumentation and other devices to measure flow properties. The modified F-106B's subsonic and transonic performance was dramatically better than the baseline airplane. (Photo courtesy of NASA)

CHAPTER 8

THE NEW MILLENNIUM

As world events transitioned into the new millennium in 2000, NASA's aeronautics program was dramatically reshaped by events and decisions made within the agency, as well as powerful external factors. Priorities assigned by NASA to the International Space Station and Space Shuttle operations, as well as new visions within the Bush Administration for manned missions to return to the Moon and perhaps explore Mars resulted in severe constraints in both manpower and funds for aeronautics research.

At the same time, the end of the Cold War had resulted in an abrupt slowdown in advanced military aircraft programs, and numerous mergers within the civil aviation industry greatly reduced the national level of innovation and emerging advanced aircraft concepts. The net result of these inhibiting factors was a drawdown of aeronautical research at the Langley Research Center, NASA, and the entire nation. NASA's aeronautics program turned toward more fundamental, theoretical analyses, and away from more expensive vehicle-oriented projects.

Despite these formidable constraints, NASA researchers and their industry and academic partners pressed on with a number of new and exciting research projects involving radical configurations. The scope of their activity included unconventional subsonic civil transports, advanced hypersonic propulsion concepts, and a remarkable aircraft design for the exploration of Mars.

The Blended Wing Body

In 1988, the NASA Langley Research Center hosted a major meeting of aeronautical engineering leaders to discuss opportunities for revolutionary advances in aircraft configurations and technology. Dennis M. Bushnell, senior scientist of the NASA Langley Research Center, challenged the group with a question: "Is there a renaissance

Artist's version of an 800-passenger Blended Wing Body civil transport. The elegant design offers potential advantages in performance, noise reduction, and emissions, but design challenges exist in aerodynamics, stability and control, propulsion, and structures. Stimulated by initial cooperative studies between NASA and McDonnell Douglas (now The Boeing Company) in the 1990s, significant progress has been made in the technologies required to mature the concept. (Graphic courtesy of NASA)

for the long-haul transport?" Bushnell's perspective on aeronautical innovation was that truly revolutionary changes in aircraft configurations had not occurred since the introduction of the B-47 bomber. The B-47 first flew on December 17, 1947 and introduced several radical features, including a swept wing and turbojet engines mounted to pylons beneath and forward of the wing. Although significant technological advancements had subsequently occurred in various aeronautical disciplines, the fundamental external geometry of aircraft in the civil-aviation jet transport fleet still retained the features of the B-47. The design practice for conventional "tube and wing" configurations had, of course, evolved over the years as the most practical and efficient approach for advanced aircraft. Bushnell and his peers, however, held on to the belief that fresh radical approaches to configuration layout might provide a new level of efficiency with significant advantages over the conventional design.

Robert H. Liebeck of the Long Beach, California, division of the McDonnell Douglas Corp., now The Boeing Company, was a close associate of Bushnell, sharing his interest in innovation and advocacy for revolutionary configurations. Stimulated by participation in the technical meeting at Langley and his ongoing personal discussions with Bushnell, Liebeck initiated a brief company study of a flying-wing type configuration that used adjacent pressurized passenger tubes aligned in a lateral plane and joined with a wing in an arrangement that vaguely resembled a tadpole. Liebeck and his McDonnell Douglas engineering team concluded that, when compared to a conventional configuration, this "blended wing" design was significantly lighter, had a higher lift-drag ratio, and had a substantially lower fuel burn. From this rudimentary design, the interest in a refined blended wing body (BWB) concept quickly accelerated.

The BWB configuration offers maximized overall aerodynamic efficiency with minimal drag, resulting in significantly improved fuel economy with reductions in undesirable emissions, and potentially lower passenger-seat-mile costs. By placing the engines above and behind the wing's upper surface, researchers predict reduced community noise during takeoff and landing

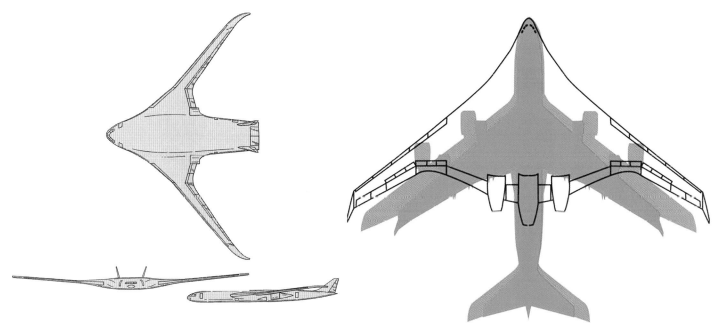

Sketch of initial BWB design by McDonnell Douglas used in preliminary estimates of aerodynamic performance. From this rudimentary beginning, the BWB configuration rapidly evolved. Numerous design issues were addressed, including aerodynamic performance, stability and control, integration of structural and propulsion systems, and operational issues such as passenger arrangement and emergency-evacuation capability. (Graphic courtesy of The Boeing Company)

Plan-view comparison of an 800-passenger BWB, the Airbus A380 (orange), and the Boeing 747-400 (green) to the same scale. By combining a two-deck arrangement with spanwise passenger seats along the thick inner wing, the BWB can efficiently carry a large payload. Note that the wingspan of the BWB is not dramatically larger than the span of a conventional transport. (Graphic courtesy of NASA)

operations. Initially, joint studies between McDonnell Douglas and NASA Langley focused on very large BWB civil transports (800 passengers), which could carry almost twice the number of passengers of a Boeing 747-400 while using advanced aerodynamic, propulsive, and structural technologies to burn only one-third the fuel for comparable operations.

At the outset, it was recognized that a BWB configuration required extensive research and development in aerodynamics, stability and control, structures and materials, and propulsion integration to reach a level of maturity and feasibility. Thus, the McDonnell Douglas-Langley team began more than 15 years of detailed studies and analyses to raise the concept's technology-readiness level. Examples of some of the issues included how to design the thick wing center body of the BWB transport for an acceptable level of aerodynamic drag, particularly at high subsonic speeds. Stability and control problems such as longitudinal instability ("pitch up") and providing adequate directional stability and control for this tailless airplane concept were anticipated based on problems encountered for flying-wing aircraft configurations in the past.

Propulsion issues were created by the possibility of undesirable inflow conditions for large turbofans in proximity to the upper wing surface. Shock waves on the wing at cruise speeds might result in poor propulsive efficiency. But perhaps the most important issue to be addressed by designers of a BWB is the structural design of the noncircular pressurized cabin. A vast amount of experience has been gained through the years with circular fuselage shapes, yielding a thorough understanding of trade-offs between weight and structural integrity, but virtually no experience had been obtained for non-circular shapes. All of these problems had to be resolved with wind-tunnel tests, computational methods, analyses of aircraft performance, and piloted simulation.

In 1994, NASA headquarters awarded a three-year project on BWB research to a team composed of McDonnell Douglas (team leader), NASA Langley, Stanford University, NASA Lewis (now NASA Glenn), the University of Southern California, the University of Florida, and Clark-Atlanta University. One of the highlights of the study was the fabrication and flight testing of a large-scale, remotely-controlled model of a 17-foot wing span BWB configuration by a team at Stanford. Powered by a pair of pusher propellers, the BWB model included an onboard computer that recorded flight-test

parameters and provided stability augmentation and control laws. In July 1997, the model was flown before an audience of industry, NASA, and DoD representatives at El Mirage, California.

When McDonnell Douglas merged with Boeing in August 1997, Boeing redirected the BWB studies toward a smaller, 450-passenger (240-foot wingspan) airplane. Also, certain configuration features, such as the propulsion installation, changed and the engines were mounted on upper surface pylons. Meanwhile, the ongoing cooperative studies on critical technologies continued between Boeing and Langley. Five different Langley wind tunnels were used for BWB tests, including the full-scale tunnel, spin tunnel, NTF (National Transonic Facility), 14- by 22-foot subsonic tunnel, and the 12-foot low-speed tunnel. The scope of tunnel testing since 1997 included evaluations of low-speed stability, control, and takeoff and landing characteristics. High-speed aerodynamic performance was also evaluated, and experimental results were correlated with computer predictions to validate design tools and provide the necessary database for further risk reduction. Today's powerful computer codes have contributed substantially to the sophistication of the BWB wing airfoil characteristics. Flying wings typically achieve longitudinal trim by using swept-back wings and downloading the wingtips, resulting in a severe

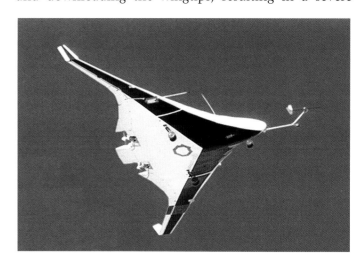

In-flight photograph of a 17-foot wingspan remotely controlled BWB model designed, built, and flown by Stanford University students and staff. Note the pusher props and the instrumentation nose boom used for in-flight measurements of flight data for analysis and feedback for stability-augmentation systems. A highly successful series of flight tests was conducted at El Mirage, California, in 1997 for representatives of industry, NASA, and DoD. (Photo courtesy of NASA)

induced-drag penalty. However, the use of advanced computational wing design methods permitted the BWB to be trimmed by a careful distribution of wing trailing-edge camber and by the judicious use of wing twist (washout).

In addition to this aerodynamic development research, Boeing and Langley have conducted exhaustive studies to design a center body for the BWB that would have satisfactory structural weight, passenger accommodation, and pressurization within the outer aerodynamic lines. This challenge is especially difficult because of the non-circular fuselage shape. In a conventional circular fuselage section, a thin skin carries internal pressure efficiently via hoop tension. If the fuselage section is noncircular, however, internal pressure loads also induce large bending stresses. Many candidate structural concepts were assessed, including honeycomb sandwich-shell and ribbed double-wall shell construction.

In yet another study, researchers conducted acoustic testing of a BWB model in an anechoic noise research facility at Langley. With its engines mounted above and to the rear of the wing, the BWB may offer significant noise shielding during near-airport operations. A high-frequency wideband noise source was placed inside the engine nacelles of the model to simulate broadband engine noise. When the research team made noise meas-

urements using a microphone array at various locations below the model, they found that the BWB configuration provided significant shielding of the simulated engine inlet noise. In fact, the noise level radiated downward into the forward sector was reduced by as much as 20- to 25-dB at certain full-scale frequencies.

In 2000, NASA initiated a new program to design, fabricate, and fly a remotely piloted sub-scale model of the BWB. Langley was assigned responsibility for the model's

BWB model without engine nacelles in the Langley National Transonic Facility during tests at high subsonic speeds for correlation with computational predictions of cruise performance. The center body lines' geometry was designed with the aid of computational fluid dynamics. The results of this particular test were significant, as the data showed that a configuration with a relatively thick center body could be designed for efficient cruise at high subsonic speeds (Mach 0.85). (Photo courtesy of NASA)

A BWB model is tested in the Langley 14- by 22-Foot Tunnel to determine recovery capability from unusual attitudes. Because some flying-wing configurations of the 1940s had exhibited longitudinal instability and tendencies to tumble nose over tail, wind-tunnel tests were designed to investigate the characteristics of the BWB design. In this photograph, the direction of the tunnel air stream is from left to right, and the model is at a large, nose-down inverted attitude to obtain data for analysis of its behavior at large negative angles of attack. (Photo courtesy of NASA)

Free-spinning test of a BWB model in the Langley 20-Foot Spin Tunnel. During the test program, researchers checked out the configuration's spin and tumbling characteristics. (Photo courtesy of NASA)

design and fabrication, with the Dryden Flight Research Center supporting development of the flight-control system. The BWB low-speed vehicle (LSV) was based on the geometry of a full-scale, 450-passenger design from Boeing studies. The LSV was designed for a wingspan of 35 feet, a maximum weight of about 2,500 pounds, and three jet engines of the 200-pound thrust class. The vehicle was given the formal designation of X-48A. Unfortunately, NASA terminated the LSV Program in 2001 because of higher-priority program commitments.

After cancellation of the X-48A, Langley redirected its interest to lower-cost, low-speed, free-flight tests of a smaller model in the Langley Full Scale Tunnel. In 2002, Boeing contracted with Cranfield Aerospace Limited in England for the design and production of a pair of 21-foot-span BWB vehicles that would ultimately be known as the X-48B.

In 2005, free-flight model tests of a 0.05-scale BWB model were conducted in the full-scale tunnel. In these highly successful flights, the 12-foot-wing-span BWB model was subjected to low-speed, high angle-of-attack conditions.

While continuing its close working relationship with Langley, Boeing has pursued other potential markets for the BWB, including potential applications as a large cargo transport or a military in-flight refueling transport. Many believe that the first applications of the BWB configuration will be for military uses, with civil applications to follow.

During 2006, BWB activities intensified, as testing of the first X-48B at Langley took place and the X-48B flight articles were readied for research flights at NASA Dryden. The Boeing Phantom Works organization led the efforts, with participation by NASA and the U.S. Air Force Research Laboratory. The Air Force involvement is an indication of the military's interest in the potential of the BWB as a long-range, high-capacity military aircraft. The first flight of the X-48B, Ship 2, at Dryden occurred on July 20, 2007, followed by additional flights and testing through the summer. After assessments of the low-speed characteristics and validation of the

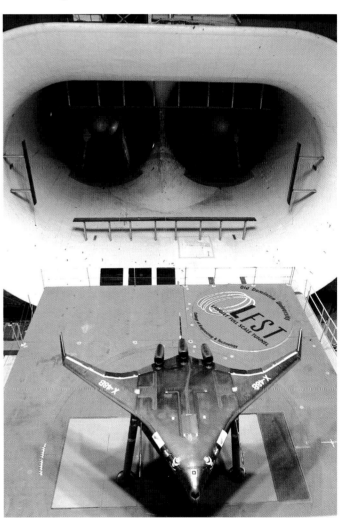

Boeing X-48B Ship 1 mounted in the Langley Full-Scale Tunnel for measurement of aerodynamic forces and moments in May 2006. Boeing Phantom Works partnered with NASA and the U.S. Air Force Research Laboratory (AFRL) at Wright Patterson Air Force Base, Ohio, to explore the advantages of the BWB configuration. After testing was completed, Ship 1 was moved to the Dryden Flight Research Center to serve as the backup for a second X-48B prototype. (Photo courtesy of NASA)

Free-flight testing of an X-48 model in the Langley Full-Scale Tunnel during an evaluation of dynamic stability and control in 2005. The 12-foot span model was also used for conventional wind-tunnel force measurements to generate aerodynamic data for the configuration. (Photo courtesy of NASA)

The 21-foot span Boeing X-48B Ship 2 at Dryden in October 2006 in preparation for flight tests. Note the tail stinger for an emergency spin parachute and the wing booms for air-data measurements. Dryden's staff has extensive experience in conducting remotely piloted tests of radical aircraft, and the test ranges at Edwards are ideal for experimental test flights. (Photo courtesy of NASA)

The X-48B on its fifth research flight at Dryden in August 2007. Objectives of initial X-48B flights were to check out aircraft systems, verify control laws, and evaluate low-speed high-lift characteristics. Note the leading-edge slats extended for the high-lift configuration. (Photo courtesy of NASA)

design-control laws for the aircraft, the X-48B team will conduct evaluations at high subsonic speeds.

Hyper X

For more than 40 years the NASA Langley Research Center has conducted research on technologies required for hypersonic (faster than Mach 5.0) air-breathing vehicles, with a special emphasis on developing a revolutionary space-launch capability for the United States. Using unique NASA hypersonic wind tunnels and advanced computational methods coupled with systems-level analysis of space-launch concepts, the center has maintained a leadership role in the development of advanced hypersonic vehicles. As the efforts continued, researchers concluded that a single-stage-to-orbit (SSTO) space transportation system was feasible with the development of required technologies. SSTO vehicles powered by scramjet engines could offer significant advantages over conventional rocket-powered space vehicles by providing an increased payload through a reduction in on-board propellant. Other advantages cited for the scramjet vehicle include increased operational flexibility, reduced costs, and increased safety.

A critical element for this research effort has been the development and maturation of an air-breathing, super-sonic-combustion ramjet, or scramjet. A scramjet is designed to compress onrushing hypersonic air in a combustion chamber. Liquid hydrogen is then injected into the chamber, where it is ignited by the hot compressed air to create thrust. Conventional rocket engines are powered by mixing fuel with oxygen, both of which must be carried onboard the vehicle. The scramjet-powered vehicle would carry only hydrogen fuel, and the atmosphere would provide the oxygen needed to burn the fuel. By eliminating the need to carry oxygen, future hypersonic vehicles could have room to carry greater payloads.

The performance of the scramjet engine is dependent on the aerodynamics of the airframe (the underside of the vehicle must function as the air inlet mechanism) and the exhaust nozzle. Careful blending and integration of the airframe and scramjet engine is critical to successful performance of the concept. Simple in theory, but exceedingly difficult to design and demonstrate, a scramjet engine has no moving parts. The main challenge for a scramjet is to introduce fuel, ignite it, and achieve combustion in the millisecond

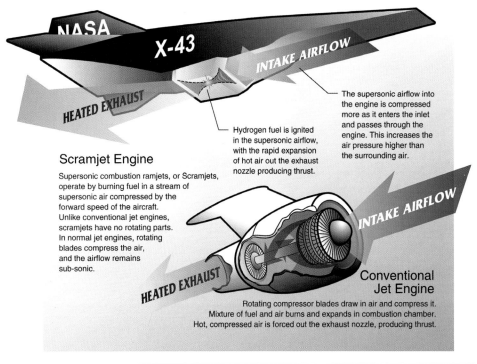

The supersonic combustion concept of the scramjet compared to a conventional subsonic turbo-fan concept. By eliminating the requirement of carrying on-board oxygen, the scramjet offers more payload capability than rocket-powered vehicles. (Graphic courtesy of NASA)

Scramjet Engine

Supersonic combustion ramjets, or Scramjets, operate by burning fuel in a stream of supersonic air compressed by the forward speed of the aircraft. Unlike conventional jet engines, scramjets have no rotating parts. In normal jet engines, rotating blades compress the air, and the airflow remains sub-sonic.

Hydrogen fuel is ignited in the supersonic airflow, with the rapid expansion of hot air out the exhaust nozzle producing thrust.

The supersonic airflow into the engine is compressed more as it enters the inlet and passes through the engine. This increases the air pressure higher than the surrounding air.

Conventional Jet Engine

Rotating compressor blades draw in air and compress it. Mixture of fuel and air burns and expands in combustion chamber. Hot, compressed air is forced out the exhaust nozzle, producing thrust.

that each individual molecule of air spends in the engine. The task makes lighting a match in a hurricane seem easy by comparison.

Since the mid-1960s, Langley has pursued research on hydrogen-powered and cooled scramjet engines, including fixed- and variable-geometry engine concepts. Military support for Langley's activities has waxed and waned through the years. However, in the early 1980s, DARPA initiated a project called Copper Canyon to develop a single-stage-to-orbit (SSTO) hypersonic airplane that could take off and land horizontally. NASA was an active participant in the effort, which lasted until 1985. The scope of Langley activities included wind-tunnel tests, systems analysis, research on structures and materials, scramjet testing with additional consultation and design, and piloted simulator studies. The focus of the project was the X-30 National Aerospace Plane (NASP), with funding provided by NASA and DoD, and with McDonnell Douglas, Rockwell International, and General Dynamics competing to develop the technology for the SSTO vehicle. Rocketdyne and Pratt & Whitney also competed to develop the engines for the X-30. Under this effort, no commitment was made to design, fabricate, or flight test an X-30 aircraft. Efforts to develop technologies continued under the leadership of Rockwell until 1993, when the NASP activity was terminated due to rapidly escalating costs, lack of a clear civil or military mission, and immature technologies.

Artist's concept of the X-30 National Aerospace Plane. The X-30 configuration was a scramjet-powered hypersonic design. The shovel-shaped forward fuselage was designed to generate a shock wave to compress air before it entered the scramjet engine, while the aft fuselage expanded the scramjet's exhaust. Started in 1990, the high-risk program was terminated in 1993 in recognition of immature critical technologies and substantial technical and funding challenges. (Graphic courtesy of NASA)

A 0.10-scale model of a NASP configuration mounted in the Langley 16-Foot Transonic Dynamics Tunnel for flutter testing in 1992. The model has been constructed with modules of structure carefully designed to replicate the dynamic characteristics of a full-scale vehicle. This tunnel has been used for more than 50 years by the aircraft industry, NASA, Department of Defense, and university researchers for clearing new designs from flutter and general research on aeroelasticity. (Graphic courtesy of NASA)

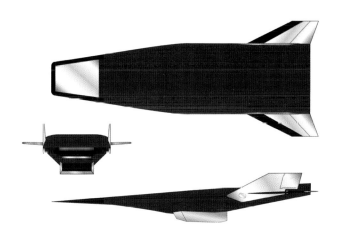

Three-view sketch of the X-43 Hyper-X research configuration. Three 12-foot-long, 5-foot-wide remotely piloted X-43 vehicles were designed and fabricated for research flights at Mach numbers of either 7 or 10. The copper-colored pod on the bottom of the vehicle is the scramjet engine. Flights were conducted at the NASA Dryden Flight Research Center in 2001 and 2004. (Graphic courtesy of NASA)

Despite the cancellation of NASP activities, interest continued in developing air-breathing hypersonic engines, and in accelerating the development of design methods, enhancing experimental facilities and databases, and validating design methodology. It was widely recognized that the integration of a scramjet-powered hypersonic aircraft would be difficult to achieve using ground-based tools, and that flight tests of a scramjet configuration would be required to reduce risk and to encourage confidence in the concept.

NASA initiated a program to demonstrate and validate the technology, experimental techniques, and computational methods and tools for design and performance predictions of hypersonic aircraft with advanced scramjet propulsion systems. Known as the Hyper-X Program, the effort focused on the development and demonstration of critical scramjet engine technologies, using three small, relatively low-cost, flight-demonstrator vehicles that would undergo flight tests at Mach 7 and Mach 10. NASA Headquarters approved the Hyper-X Program in July 1996, and it officially started in September 1996. Langley became the lead center, with responsibility for program management and technology applications. Dryden became the lead center for flight research, with the added responsibility for managing contracts with ATK GASL (formerly MicroCraft, Inc.) for fabrication of the three X-43A flight research vehicles, fuel systems, and scramjet engines; Orbital Sciences Corp. for the modified Pegasus first-stage launch vehicles that would propel the X-43 vehicles to test conditions; and Boeing Phantom Works, which provided the avionics system and final structural design.

The Hyper-X test configuration, designated the X-43, was a 12-foot-long research vehicle that would provide flight data for a hydrogen-fueled, airframe-integrated scramjet engine. In addition to fundamental information on engine performance, associated data on aerodynamic, thermal, structural, guidance, and flush-air-data-systems would be obtained. Flight data obtained would be used directly to validate design methodology for hypersonic vehicles. On August 11, 1998, the first piece of X-43 hardware was delivered to NASA in the form of a scramjet engine that was used for a series of ground tests in Langley's 8-Foot High-Temperature Tunnel.

The flight plan for the unpowered X-43 vehicle was based on a launch from a NASA B-52 mother ship at an altitude of 19,000 feet and a Mach number of 0.5. After

Pegasus/X-43 model mounted in Langley 20-Inch Mach 6 Wind Tunnel for hypersonic tests in 1996. Inspecting the model is Vincent L. Rausch, Langley's X-43 Program Manager. The experimental results, obtained on a high-fidelity 12.5-percent scale model, indicated well-behaved characteristics for the anticipated flight conditions. (Photo courtesy of NASA)

X-43 scramjet undergoing power-on tests at Mach 7 in the Langley 8-Foot High Temperature Tunnel in 1999. This test provided engine-performance and operability data, as well as design and database verification for the Mach 7 flight tests, which resulted in the first-ever airframe-integrated scramjet flight data. The Hyper-X flight engine, a duplicate Mach 7 X-43 scramjet engine, was mounted on an airframe structure that duplicated the entire three-dimensional propulsion flow path from the vehicle's leading edge. This model was also tested to verify and validate the complete engine system. Langley researchers conducted wind-tunnel tests on more than 20 scramjet engine configurations. (Photo courtesy of NASA)

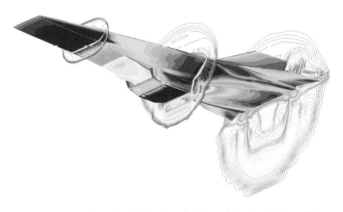

Advanced computational methods played a key role in the design of the X-43. This computer-derived simulation is for Mach 7 with the scramjet operating. The development of the Mach 7 X-43 flights required a pre-flight assessment of longitudinal and lateral-directional characteristics near the flight-test condition. The database development was accomplished through extensive correlation of wind-tunnel testing and computational analyses. (Graphic courtesy of NASA)

launch, the Pegasus booster would be ignited and, when the scramjet test condition was reached at approximately 100,000 feet, the X-43 would be pyrotechnically separated. After separation, the scramjet engine cowl inlet door was opened, and the scramjet fueling sequence commenced. A combination of silane and gaseous hydrogen was injected into the combustor region, and data would be taken. After the test sequence was completed, the vehicle followed a controlled tra-jectory as it descended and decelerated through the supersonic and transonic flight regimes before termination at subsonic conditions.

The first X-43 flight on June 2, 2001, was planned for a test condition of Mach 7, and the launch from the B-52 occurred at a Mach 0.5 at 23,000 feet. Unfortunately, at 13.5 seconds after release and at an altitude of approximately 22,000 feet, structural overload of the starboard elevon occurred. The severe loss of control caused the

Captive flight of the Pegasus/X-43 stack on the NASA Dryden B-52 just after takeoff on September 27, 2004. The flight was designed to check out systems prior to the world-record Mach 10 flight later in the year. NASA's B-52B launch aircraft participated in some of the most significant projects in aerospace history. At its retirement in December 2004 following the X-43 flights, the aircraft held the distinction of being NASA's oldest aircraft, as well as being the oldest B-52 on flying status. At the same time, it had the lowest number of flying hours (2443.8) of any B-52 in operation, having been used exclusively in the role it continued to perform so reliably in for nearly 50 years. (Photo courtesy of NASA)

Flight trajectory derived for the successful record-breaking X-43 research flights following an investigation of the first unsuccessful flight. Typically, during low-Earth orbit insertion for small payloads, the Pegasus booster is released at 40,000 feet. However, in the first X-43 flight the Pegasus was released at only 23,000 feet and the denser atmosphere doubled the aerodynamic loads acting on the tail surfaces. The increased forces acting on the fins apparently overstressed the surfaces, causing them to fail and send the Pegasus out of control. The release altitude was changed to 40,000 feet in subsequent flights. In

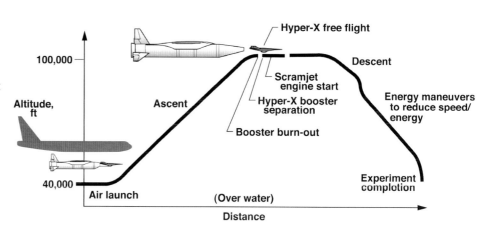

addition, the Pegasus's tail-fin actuators were redesigned with larger torque margins to handle the aerodynamic loads. (Graphic courtesy of NASA)

X-43A to deviate from its planned trajectory, and as a precaution, it was destroyed by range safety 48.6 seconds after release. The test article was lost, and a grueling two-year investigation of the cause of the accident and cures for the problem ensued. Langley assembled a huge tiger team of aerodynamicists and flight-dynamics experts to conduct appropriate wind-tunnel tests and flight-control analyses. The cause of the accident was ultimately identified as inaccurate pre-flight estimates of the interactions between the control system and the transonic aerodynamic loading for the launch conditions. As a result of

these findings, the launch speed and altitude for future flights were increased to Mach 0.8 and 40,000 feet.

The highly successful flight of the second X-43 vehicle occurred on March 27, 2004, for a Mach number of 7. Launched from the new initial conditions of Mach 0.8 and 40,000 feet, the Pegasus ignited after a five-second free fall to about 39,500 feet. The stack executed a pull-up, followed by a pushover to an altitude of 95,000 feet. Following booster burnout, stage separation, and X-43 vehicle stabilization, the engine inlet opened for about 30 seconds of testing and data

Pegasus rocket ignites during the record-setting X-43 flight of November 16, 2004. Guinness World Records recognized NASA's X-43A flight with a new world speed record for an air-breathing aircraft: Mach 9.6, or nearly 7,300 mph. The X-43A set the new mark and broke its own world record on this third and final flight of the program. (Photo courtesy of NASA)

acquisition. After the data was acquired, the engine cowl closed, and the vehicle flew a controlled descent to "splash-down" in the Pacific Ocean. The record-breaking flight was the highest speed attained by an air-breathing aircraft, and the measurement of positive acceleration at the hypersonic condition was a remarkable breakthrough for scramjet technology.

The third X-43 flight, on November 16, 2004, was an overwhelming success and gained world-wide attention by setting a speed record. With a planned test condition of Mach 10, the challenges to the engineering team were significantly increased. For example, the peak heating loads on the vehicle at Mach 10 would be twice those encountered in the previous Mach 7 flight. Following the same launch conditions, the "stack" executed a pull-up to a flight path angle of more than 30 degrees, followed by a push-over to achieve flight at an altitude of 110,000 feet. After booster burnout, stage separation, and stabilization of the X-43 vehicle, the engine was opened for about 20 seconds, including 11 seconds of powered flight at Mach 9.68, and additional data was gathered as the vehicle slowed down. Prior to the flight, the total amount of X-43 wind-tunnel test data at Mach 10 (gathered a few milliseconds at a time) totaled no more than one to two seconds. Thus, the database had increased by an order of magnitude. Once again, positive acceleration during the powered phase of flight verified the performance of the scramjet engine. This third X-43 flight was the fastest ever by an air-breathing vehicle. Currently, the world's fastest air-breathing aircraft, the SR-71, cruises slightly faster than Mach 3. The X-43's flight speed was even faster than the legendary rocket-powered X-15, which reached its highest speed of Mach 6.7 in 1967.

The X-43 research vehicles' highly successful flight tests have provided the hypersonic community with

critical demonstrations of the capability of hypersonic air-breathing vehicle design tools. The Hyper-X Program was a small but crucial step toward maturing air-breathing hypersonic propulsion for application to space-launch vehicles and other efficient hypersonic systems. Despite these revolutionary accomplishments, it is obvious to designers that many flight tests must be performed in conjunction with wind-tunnel tests, systems analyses and engineering, and basic research, before scramjet technology will be ready for practical applications to future vehicle systems.

Mars Airplane

The concept of using an airplane to explore and conduct science missions in the atmosphere of the planet Mars has existed in literature for over half a century. This exciting idea has been formulated on the theory that airborne observations of Mars would have several advantages that can complement other exploratory methods, such as robotic land rovers and orbiting satellites. By their very nature, land rovers conduct close-up, microscopic examinations of terrain and terrestrial features, but are limited in their range and relative coverage. On the other hand, satellites map out global areas, but with relatively low resolution. Spanning these capabilities, an airplane flying at relatively low altitude (approximately one mile) could carry instrumentation for high-resolution sampling over a relatively large distance (about 400 miles), and permit navigation to specific areas of interest.

During the last 35 years, the use of remotely piloted, unmanned aircraft for the exploration of Mars has been proposed in paper studies and engineering analyses. Such airplane concepts typically involve transporting an unmanned aircraft and its propulsion

Artist's concept of a fixed-wing science platform airplane flying over Mars. Exploratory low-level flights in the Martian atmosphere would complement missions of robotic rovers and orbiting satellites, while providing more flexibility in data accuracy and scope. Long a dream of scientists, the Mars-airplane concept faces many technology hurdles. (Graphic courtesy of NASA)

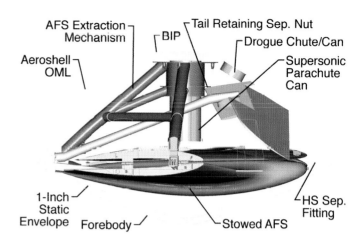

Conceptual packaging of Mars airplane in an aeroshell capsule. Transport of a Mars airplane by an interplanetary rocket is a major challenge because constraints on vehicle size result in vehicle complexity. Note the stowed aircraft flight system, the folded tail and wing structures, the aircraft extraction system, parachutes for deceleration and stabilization of the aeroshell and the deployed aircraft, and the forebody/heat shield. (Graphic courtesy of NASA)

system in an enclosure on an interplanetary rocket to the vicinity of the Martian atmosphere. To accomplish this task, the aircraft would necessarily be of variable geometry, or foldable, and contained in an aeroshell for protection during atmospheric entry. At a specific altitude above the Martian surface, the aeroshell would be released and the aircraft deployed, unfolded, and maneuvered under power to perform the exploration mission.

The NASA Jet Propulsion Laboratory had sponsored such a study in the late 1970s, which proposed a complex, propeller-driven folding airplane with multiple wing folds. This vehicle used six wing folds, three fuselage folds, and a folding propeller. The results, together with those of other past NASA, industry, and university studies, identified some of the numerous challenges and risks facing the successful design of such a vehicle, as well as the design parameters and concepts that might be used for the configuration.

Some of the major issues to be addressed include: the geometric constraints on airframe dimensions; the fact that the aircraft will be in a dormant, stored configuration for a long period of time during the transit to

Mars (about a year); thermal, acceleration, and radiation effects; and structural loads during launch and entry while the airplane is folded. The flight environment for the mission will be especially challenging. The Martian atmospheric density is one one-hundredth of Earth's at sea level, and the lack of oxygen in the atmosphere requires the use of inefficient non-air-breathing propulsion. In addition to being about 100 degrees colder than Earth, the speed of sound in the Martian atmosphere is lower, resulting in undesirable aerodynamic compressibility effects at lower flight speeds than on Earth. Finally, critical issues exist in the areas of guidance, navigation, and control. For example, autonomous flight is required because of the long time delay in control signals, which, although traveling at the speed of light, must be transmitted over vast distances. If an operator on Earth saw data from the aircraft indicating that a flight input needs to be made, that information is already three- to 22-minutes old, depending on the orbital alignment between Earth and Mars. Then, the pilot's corrective inputs would take just as long to reach the aircraft flying above the Martian landscape. That's like driving down the street, turning the steering wheel, and having the automobile turn many minutes later. In addition, navigational aids such as GPS will not exist, and there is no global magnetic field. Another major problem to be faced during

Sequence of events from entry to deployment of aircraft and research flight. (a) Aeroshell enters Martian atmosphere during entry, descent, and deployment sequence. (Graphic courtesy of NASA)

(b) Parachute inflates and heat shield is released. (Graphic courtesy of NASA)

(c) Airplane is ejected from aeroshell. (Graphic courtesy of NASA)

(d) Airplane unfolds and deploys for pull-up maneuver. (Graphic courtesy of NASA)

(e) Airplane stabilizes and releases drogue chute. (Graphic courtesy of NASA)

(f) Scientific mission initiated.

precision flights will be the uncertainty in environmental features, such as winds and local topography.

In the 1990s, NASA developed an interest in a Mars Micromission Project with the objective of sending a series of small, low-cost spacecraft to Mars on Arianne 5 rockets for science missions. In February 1999, NASA Administrator Daniel Goldin stimulated a potential project to fly an unmanned aircraft on Mars in conjunction with the 100th anniversary of the Wright Brothers' flight at Kitty Hawk, North Carolina, on December 17, 2003. A large team of NASA personnel from several centers was quickly formed to begin the formidable task of designing and maturing a flightworthy airplane for an Arianne launch vehicle. Langley was assigned overall responsibility for the project, known as the Mars Airplane Package (MAP). The work included the development of the entire system that would enter the Martian atmosphere, including the

aeroshell atmospheric entry system, the variable-geometry airplane, and the scientific instruments. The aeroshell container for the aircraft had a diameter of only 2.6 feet, forcing designers into adopting a very small airplane with a wingspan of only 5.6 feet. Because of its small size, the proposed vehicle's performance was severely limited, and the project was viewed within the serious scientific community as more of a "stunt" than a true science platform. The Mars Airplane Project was cancelled in November 1999, but the NASA teams at the Langley, Glenn, and Dryden centers had laid the groundwork and gained experience for the design of more feasible airplane configurations and system concepts to permit atmospheric flights on Mars.

At Langley, Dr. Joel S. Levine, an internationally recognized expert in atmospheric sciences, was the leader and chief advocate for a Mars airplane activity. Levine had participated in the MAP study, but he wanted a true science-focused mission that represented significant value to the scientific community, as well as one that would also provide critical data for potential landing sites to the human spaceflight segment of NASA. Alert to expressions of mounting interests in Mars exploration by NASA Headquarters in 2000, Levine initiated several in-house and contractual efforts to prepare Langley's competitive position to respond to possible opportunities to initiate a Mars airplane research program. Internally, staff members from various areas of expertise within Langley joined the team with representation in the disciplines of aerodynamics, structures and materials, atmospheric sciences, and system studies. A broad range of exploration vehicles, including rovers, orbiters, balloons, airships, gliders, and powered airplanes, were studied for the Mars mission. Only the powered airplanes would meet the goals set by the team. Since the Martian atmosphere is so thin, propeller-driven designs must use very large-diameter props, resulting in aerodynamic issues such as tip Mach number and airfoil challenges. Folding propeller designs introduced even more complications. After in-depth trade studies, the team chose a bi-propellant rocket propulsion system. The airframe concept was a foldable configuration with an inverted-V tail mounted on twin booms to simplify the folding procedure and to avoid the high temperatures of the rocket exhaust.

Exploratory wind-tunnel studies were initiated to examine aerodynamic trades and assess the complications of deploying a foldable aircraft. In March 2002, wind-tunnel tests were conducted in the Langley 16-Foot Transonic Dynamics Tunnel (TDT) to evaluate wing deployment performance for a wide range of vehicle attitudes and aerodynamic conditions. The TDT is a special, variable-pressure wind tunnel that allowed the research team to match the Mach number and Reynolds number conditions that the airplane would experience on Mars. These were dynamic tests, in which the folded wing panels were released and allowed to swing into their deployed and locked positions.

In January 2002, Langley contracted Aurora Flight Sciences Corp. of Manassas, Virginia, to support the development of a possible proposal by demonstrating the feasibility of deploying a sub-scale aircraft at conditions representative of the thin Martian atmosphere. The NASA/Aurora team planned for flight tests of two aircraft. The first vehicle was known as the High-Altitude Deployment Demonstrator 1 (HADD1), which repre-

sented a 50-percent scale version of a full-scale aircraft configuration. The second vehicle, known as HADD2, was a full-scale design. Design, construction, and flight testing of HADD1 began at a frantic pace. Bench testing of overall system compatibility, deployment testing, and environmental testing were conducted in preparation for flight tests. The aircraft was first evaluated in a low-altitude test flight in July 2002, during which the folded airframe was lifted by a manned hot-air balloon at a test site in West Virginia and released from an altitude of 2000 feet. The deployment, pull-out to controlled flight and return to a grass runway were conducted in an autonomous operation, and then control of the vehicle was taken over by a human pilot via remote control for the successful landing. After the learning experiences of the low-altitude test, the vehicle was prepared for its high-altitude drop test at a site in Oregon later in the year.

On May 1, NASA's Office of Space Science announced the opening of competition within the scientific community for a program known as Mars Scout. Mars Scout investigations were science missions, led by scientists such as Levine and his peers, and intended to complement and extend the established NASA Mars Exploration Program. The plan called for the selected Mars Scout mission to be launched in 2007. With considerable homework and proposal data in hand, Levine and his team submitted a proposal in August 2002 entitled, "Aerial Regional-scale Environmental Survey (ARES) of Mars." Levine was identified as the principal investigator for the team. In its submittal, Langley proposed a robotic, rocket-powered, controlled airplane as a revolutionary new platform for the scientific exploration of Mars. The proposed vehicle mission would, for the first time, explore the Southern Highlands of Mars.

Meanwhile, the HADD1 demonstration aircraft had been prepared for a critical high-altitude test, designed for conditions deemed more representative of a Martian deployment. Attached to a high-altitude helium balloon on September 19 and carried to an altitude of approximately 100,000 feet over Oregon, the test airplane dropped from the balloon, unfolded, and completed a 90-minute, pre-programmed flight path. After being released over the Pacific Ocean and flying autonomously for nearly 200 miles, the 50-pound, 10-foot wingspan aircraft was flown by a human pilot during the last few minutes of flight and landed safely at the runway from which it had been launched.

In December 2002, NASA selected the Langley ARES proposal as one of four final candidates for the 2007 Scout mission. The four selected proposals were judged to have the highest science value among 25 proposals submitted to NASA. Following detailed mission-concept studies by the competitors, a NASA selection board planned to recommend one of the proposals for development and flight in 2007. Led by Langley, ARES was composed of a team of industry, academia, and national laboratories that had been actively working to prepare aircraft technology for a scientific application on Mars.

Aurora Flight Sciences/NASA flight-test team poses with the HADD1 high-altitude demonstration vehicle. Launched from a facility at Tillamook, Oregon, on September 19, 2002, the unpowered vehicle was folded and carried to an altitude of about 100,000 feet by a high-altitude balloon and released, followed by a highly successful deployment, extended flight, and recovery landing. (Photo courtesy of NASA)

Excitement among the team members rapidly escalated as it submitted its conceptual study report to NASA and conducted a site visit by selecting officials. The team viewed ARES as a major expansion of previous Viking, Mars Global Surveyor, Odyssey, and Mars Pathfinder missions, providing a window into the structure and evolution of the atmosphere, surface, and interior of Mars. If selected, ARES would be launched in September 2007 on a Delta 2925 launch vehicle and would fly through the atmosphere of Mars in September 2008.

In the ARES concept, the airplane is delivered to Mars in an aeroshell attached to a carrier spacecraft. The aeroshell separates from the carrier approximately nine hours prior to entry into the Martian atmosphere. The spinning aeroshell, which is unguided, decelerates in the Martian atmosphere and is followed by a parachute deployment to further reduce its speed. The heat shield of the aeroshell is released, followed by the release of the airplane. At an altitude of about 26,000 feet and a Mach number of 0.5, the airplane is released, unfolded, completes a pullout maneuver, and then begins level flight using its rocket motor. Continuous communication occurs between the airplane and the flyby ARES carrier spacecraft until the end of the flight.

ARES had three major science goals, including measuring and understanding the detailed nature of crustal magnetism on Mars; measuring and improving the understanding of the near-surface atmospheric composition; and providing an improved understanding of water-equivalent hydrogen abundance and its relationship

The past meets the future. The HADD1 in the pre-flight mist faces a venerable DC-3 on the morning of the high-altitude launch test. (Photo courtesy of NASA)

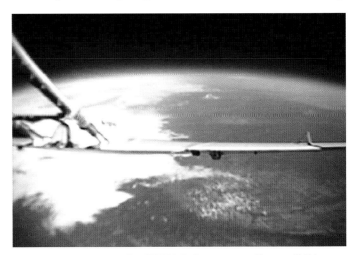

View from tail camera during HADD1 deployment over Oregon. At this point the wings and tail have unfolded, the tail-drogue parachute has been cut loose following stabilization of the vehicle, and the pullout maneuver has been completed. (Photo courtesy of NASA)

to inferred near-surface water and hydrated minerals. From its unique vantage point about a mile above the surface of Mars, ARES would target and explore up to about 380 miles of diverse terrain in the Southern Highlands. Science data would be returned to Earth starting on the day of flight for immediate scientific review and public dissemination.

In August 2003, NASA completed its evaluations of the four competing proposals and selected the Phoenix Mars Lander, proposed by the University of Arizona's Lunar and Planetary Laboratory, for the first Mars Scout mission in 2007. Phoenix used a stationary robotic lander spacecraft to search for environments suitable for microbial life on Mars, and to research the history of water there. Phoenix was landed in the water-ice-rich northern polar region of Mars in May 2008.

Despite the disappointing loss of opportunity in 2003 for the ARES concept and the impact on morale, the Langley-led team continued to assess and mature the technologies required for a more robust flyer design and reduce the technical risk. Ejection dynamics tests were conducted in 2004. Also, the TDT was used for tests of the refined ARES design.

Efforts were put into even more refinements to the ARES design, leading to an ARES-2 configuration. The three primary areas of emphasis in the refinement were the wing airfoil, fuselage shape, and tail geometry. A new airfoil distribution across the wing was adopted, resulting in decreased trim drag and better lift-to-drag ratio for improved performance. The fuselage aerodynamic characteristics were dramatically improved by re-orienting the fuel tanks in a transverse, rather than longitudinal direction, permitting a mid-wing arrangement and further reducing drag. This arrangement of fuel tanks required additional analysis of the potential effects of fuel sloshing on lateral control. Studies of the capsule's spin modes were conducted, the potential for adverse aerodynamic interference effects between the aeroshell and the airplane were assessed, and deployment maneuvers were devised. Team members from the Charles Stark Draper Laboratory of Cambridge, Massachusetts, provided contributions in the design of the guidance and navigation systems for the airplane. The required flight path for scientific missions was an altitude of about one mile with a tolerance of about 500 feet. Altitude would be maintained by on/off modulation of the rocket engine over the course of the flight.

The team's participants included a broad spectrum of NASA, academia, and industry. The science team, led

Refined ARES configuration during aerodynamic testing in the Langley 16-Foot Transonic Dynamics Tunnel in May 2004. Analysis of aerodynamic data led to modifications in wing airfoil distribution, fuselage shape, and tail geometry from the earlier ARES-1 configuration. (Photo courtesy of NASA)

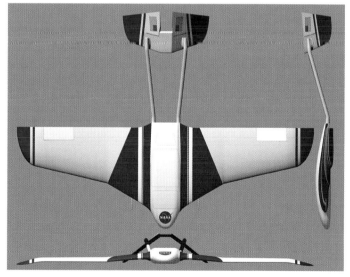

Three-view sketch of the refined ARES vehicle (ARES-2). Reconfiguration of the internal fuselage fuel tank from a single fore-and-aft container to two spanwise tanks permitted the fuselage to be configured into a more streamlined shape with less camber. In turn, the reduced camber resulted in less pitching moment and a reduction in trim drag. Lift/drag ratio was improved by more than 15 percent, greatly increasing the vehicle's performance capability. (Graphic courtesy of NASA)

by Levine, was composed of representatives from Langley, Ames, JPL, Goddard, Los Alamos National Laboratory, Arizona State University, Brown University, University of Texas at Dallas, Nantes University of France, Cornell University, and George Mason University. The ARES flight-systems team was composed of members from Langley, Goddard, JPL, Lockheed Martin Astronautics, Aurora Flight Sciences, Charles Stark Draper Laboratory, Malin Space Science Systems, and the University of Texas, Dallas.

On May 1, 2006, NASA Headquarters announced the second mission of the Mars Scout Program, which would occur in 2011, with proposals due in summer of 2006. In July, Langley submitted its ARES proposal for the Mars Scout 2006 competition amid optimism and confidence. In September, Aurora Flight Sciences delivered the full-scale HADD2 Engineering Development Model to Langley for integration of the scientific instruments and sub-systems (propulsion system, telecommunications system, etc.), and to get an early start on verifying the form, fit, and function of the assembled airplane.

On January 8, 2007, NASA had completed its review of 26 proposals and announced two winners as its selection for the 2011 missions. Known as Mars Atmosphere and Volatile Evolution (MAVEN) and The Great Escape, both projects will use satellite orbiters and will focus on the upper Martian atmosphere and ionosphere. Joel Levine was subsequently selected by NASA to serve as the Mars Scout Program Scientist.

Throughout its developmental life, the ARES concept was judged by many to be a relatively high-risk venture compared to the rover and orbiter concepts for which experience and confidence had already been attained. Maintaining a low-risk, relatively low-cost capability of value to the scientific community for exploration of Mars will continue to be the major barrier to exploratory flyers, requiring a significant commitment to further research and development. Visionaries such as Levin, however, predict that future airplanes will be used to explore not only Mars, but also Venus, Jupiter, and Saturn. First, however, a radical demonstrator similar to ARES must lead the way.

Some of the ARES research team members pose with the full-scale high-altitude drop vehicle. With a wing span of about 21 feet, the airplane provides a visual calibration of the size of a potential Mars airplane of the future. Unfortunately, the ARES project was terminated before the planned flight tests could be conducted. (Photo courtesy of NASA)

CHAPTER 9

REFLECTIONS

This collection of technical snapshots and anecdotal notes has attempted to convey some of the research activities on radical and unconventional aircraft concepts conducted by the NACA and NASA at the Langley Research Center. Hopefully, the material has brought to light many contributions and partnerships that are not generally known or documented in the literature and lore of aircraft history. It is extremely gratifying to recognize that the objectives of the founding fathers of the NACA to identify and solve the problems of flight have been successfully accomplished within a government agency. It is also impressive that old NACA wind-tunnel facilities that were put into operation to test biplanes are still operating today and conducting aerodynamic research on applications that were totally unimaginable when they were designed.

Once again, it should be remembered that the material discussed herein has necessarily been limited to just a few interesting examples of NACA/NASA research contributions in a few facilities at a single NASA aeronautics center. Furthermore, the theme of this publication has centered on radical, unconventional aircraft configurations that were not, or have

not yet been, put into production. Much, much more information is available regarding the use of NACA and NASA contributions of data, concepts, and support for well-known production aircraft. Virtually every current production U.S. military and civil aircraft flying today has made use of NASA's recent and legacy aeronautical technology, and coverage of the vast numbers of interesting design and developmental interactions far exceeds the bounds of this book. Fortunately, many references covering most of the programs are available to the reader, and excellent NASA websites containing extensive technical reports, historical photographs, and program summaries are now available for use.

Radical, unconventional aircraft designs generally capture the attention and interest of aviation enthusiasts and the public because such aircraft look different. During the past century of powered flight, aircraft configurations have evolved along traditional lines for specific applications. To the layman, subsonic commercial jet transports all appear to be of a similar swept-wing shape, high-performance military fighters all have twin vertical tails, and supersonic cruise transports have highly swept wings. Of course, different types of aircraft configurations evolve for rigorous reasons. The potential introduction of novel unconventional designs must adequately address the potential advantages of changing from standard practices—a formidable task.

For radically new designs to become practical and enter production, they must first meet and convincingly surpass technical requirements based on performance and suitability for the envisioned application. This requirement may, in itself, serve as a barrier for further development and cause the idea to fall by the wayside. But, even after technical problems have been hurdled, researchers and engineers must address a long list of potential barriers to acceptance. These issues may be grouped into considerations of engineering feasibility, economics, safety, and environmental impact. In the area of engineering, factors such as produceability, maintainability, reliability, and airport compatibility require appropriate analysis and resolution. Perhaps the most powerful barriers to production involve economic considerations such as affordability, market demand, timeliness, certification, operating costs, and resource availability. Resolution of safety issues such as structural integrity, flying qualities, failure modes, emergency egress, icing protection, and crashworthiness is mandatory. Finally, today's environmentally sensitive culture demands that unconventional vehicles comply with ever-increasing stringent regulations to protect the public and its planet. Regulatory compliance for noise levels at and near the airport and emissions into the atmosphere must be assured, not only for current regulations, but also for projected future levels of constraint. Many of the foregoing issues were key factors in the demise of the historic radical aircraft discussed herein, and serve to emphasize that just because an aircraft looks different is no assurance of a successful development program or production.

A multitude of constraints present in our world has led to a downturn in research on unconventional aircraft designs. Currently, far fewer new aircraft designs are being pursued by far fewer industry design teams, and the end of the arms race of the Cold War has stifled the introduction of new military aircraft. NASA's mission has been strongly directed at a return to the moon and beyond and away from aeronautics. Ironically, however, the U.S. public is in an almost constant outcry with complaints about poor airline service and unreliable flight schedules at a time when the projected demand for air travel is soaring. The price of fuel has skyrocketed beyond any of the most pessimistic projections of just a few years ago. Perhaps the coming years will see a re-focus of this nation's expertise and capabilities to better serve the flying public and the public in general, maybe even with advanced radical aircraft. Many naysayers proclaim that the airplane has traveled the evolutionary road, that the traditional tube-and-wing configuration will never be surpassed, and that no revolutionary improvements are possible in aeronautics.

We think not.